Introduction to Laser Technology

Introduction to Laser Technology

Deborah Clooney

CLANRYE
INTERNATIONAL
www.clanryeinternational.com

Clanrye International,
750 Third Avenue, 9ᵗʰ Floor,
New York, NY 10017, USA

ISBN: 978-1-64726-136-8

Cataloging-in-Publication Data

Introduction to laser technology / Deborah Clooney.
    p. cm.
Includes bibliographical references and index.
ISBN 978-1-64726-136-8
1. Lasers. 2. Lasers in engineering. 3. Lasers--Industrial applications. I. Clooney, Deborah.
TA1675 .I58 2022
621.366--dc23

For information on all Clanrye International publications
visit our website at www.clanryeinternational.com

LANRYE
INTERNATIONAL

Contents

Preface

A laser is a device which emits a coherent beam of light through the process of optical amplification. The basic principle behind lasers is the stimulated emission of electromagnetic radiation. The word laser is an acronym for light amplification by stimulated emission of radiation. Laser beams possess spatial coherence, which makes it possible to focus them on a very tight spot. Some of the myriad devices which make use of laser technology are laser printers, fiber-optic communication, barcode scanners, optical disk drives and welding materials. There are various types of lasers, based on the materials which are used to make them. A few major types are gas lasers, chemical lasers, excimer lasers, solid-state lasers and fiber lasers. Laser technology is an upcoming field of science that has undergone rapid development over the past few decades. Some of the diverse topics covered in this book address the varied branches that fall under this category. As this field is emerging at a fast pace, this book will help the readers to better understand the concepts of this field.

A short introduction to every chapter is written below to provide an overview of the content of the book:

Chapter 1 - A device which emits light by using a process for optical amplification is termed as a laser. It is based on the stimulated emission of electromagnetic radiation. This is an introductory chapter which will provide a brief introduction to all the significant aspects of lasers as well as its various components; **Chapter 2 -** There are numerous types of lasers. Some of them are solid state laser, dye laser, semiconductor laser, gas laser, chemical laser, copper vapor laser, argon ion laser, krypton ion laser and free electron laser. The topics elaborated in this chapter will help in gaining an extensive understanding about the diverse aspects of these types of lasers; **Chapter 3 -** Some of the diverse characteristics which are studied regarding lasers are the optical cavity, line width, beam quality, optical intensity, peak power and slope efficiency. The topics elaborated in this chapter will help in gaining a better perspective about these characteristics of lasers; **Chapter 4 -** Lasers are applied in numerous fields for a variety of purposes. Some of these are satellite laser ranging, communication, laser cutting, laser welding, scanning for building design and construction, and printing. These diverse applications of lasers have been thoroughly discussed in this chapter; **Chapter 5 -** The safe design, use and implementation of lasers for the purpose of minimizing the risk of accidents related to lasers is termed as laser safety. Some of the different types of hazards related to lasers are electrical hazards, fire hazards and explosive hazards. All the diverse safety principles related to lasers as well as these hazards have been carefully analyzed in this chapter.

Finally, I would like to thank my fellow scholars who gave constructive feedback and my family members who supported me at every step.

Deborah Clooney

Lasers: An Introduction

A device which emits light by using a process for optical amplification is termed as a laser. It is based on the stimulated emission of electromagnetic radiation. This is an introductory chapter which will provide a brief introduction to all the significant aspects of lasers as well as its various components.

Light

Light is the electromagnetic radiation that can be detected by the human eye. Electromagnetic radiation occurs over an extremely wide range of wavelengths, from gamma rays with wavelengths less than about 1×10^{-11} metre to radio waves measured in metres. Within that broad spectrum the wavelengths visible to humans occupy a very narrow band, from about 700 nanometres (nm; billionths of a metre) for red light down to about 400 nm for violet light. The spectral regions adjacent to the visible band are often referred to as light also, infrared at the one end and ultraviolet at the other. The speed of light in a vacuum is a fundamental physical constant, the currently accepted value of which is exactly 299, 792, 458 metres per second, or about 186, 282 miles per second.

No single answer to the question "What is light?" satisfies the many contexts in which light is experienced, explored, and exploited. The physicist is interested in the physical properties of light, the artist in an aesthetic appreciation of the visual world. Through the sense of sight, light is a primary tool for perceiving the world and communicating within it. Light from the Sun warms the Earth, drives global weather patterns, and initiates the life-sustaining process of photosynthesis. On the grandest scale, light's interactions with matter have helped shape the structure of the universe. Indeed, light provides a window on the universe, from cosmological to atomic scales. Almost all of the information about the rest of the universe reaches Earth in the form of electromagnetic radiation. By interpreting that radiation, astronomers can glimpse the earliest epochs of the universe, measure the general expansion of the universe, and determine the chemical composition of stars and the interstellar medium. Just as the invention of the telescope dramatically broadened exploration of the universe, so too the invention of the microscope opened the intricate world of the cell. The analysis of the frequencies of light emitted and absorbed by atoms was a principal impetus for the development of quantum mechanics. Atomic and molecular spectroscopies continue to be primary tools for probing the structure of matter, providing ultrasensitive tests of atomic and molecular models and contributing to studies of fundamental photochemical reactions.

Light transmits spatial and temporal information. This property forms the basis of the fields of optics and optical communications and a myriad of related technologies, both mature and emerging. Technological applications based on the manipulations of light include lasers, holography, and fibre-optic telecommunications systems.

In most everyday circumstances, the properties of light can be derived from the theory of classical electromagnetism, in which light is described as coupled electric and magnetic fields propagating through space as a traveling wave. However, this wave theory, developed in the mid-19th century, is not sufficient to explain the properties of light at very low intensities. At that level a quantum theory is needed to explain the characteristics of light and to explain the interactions of light with atoms and molecules. In its simplest form, quantum theory describes light as consisting of discrete packets of energy, called photons. However, neither a classical wave model nor a classical particle model correctly describes light; light has a dual nature that is revealed only in quantum mechanics. This surprising wave-particle duality is shared by all of the primary constituents of nature (e.g., electrons have both particle-like and wavelike aspects). Since the mid-20th century, a more comprehensive theory of light, known as quantum electrodynamics (QED), has been regarded by physicists as complete. QED combines the ideas of classical electromagnetism, quantum mechanics, and the special theory of relativity.

Properties of Light

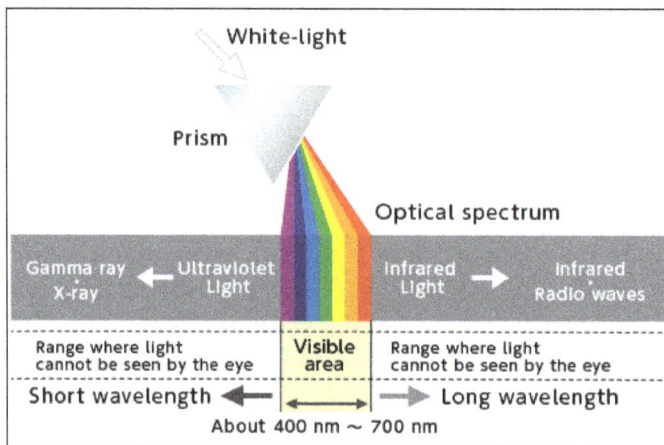

Optical spectrum.

Light has the properties of a wave and a particle. The word "wavelength" is used to express the wave or undulating property of light. It is the distance that light travels in one oscillation, and is often expressed using a unit called "nanometer". One nanometer is equal to one billionth of a meter. Our eyes can only see light that is of a wavelength between approximately 400 to 700 nanometers. This range is called the visible light. The light of other wavelengths includes X-rays, ultraviolet rays, and infrared rays. Though we cannot see them directly, these are also members of the light family.

On the other hand, light also has the property of a particle. The intensity of the light varies depending on the number of particles. Bright light has many particles while dark light has fewer particles. These particles of light are called "photons".

We can check out the particle property of light by comparing light with sound using a device called oscilloscope. Sound is known to have the characteristics of a wave. When the intensity or magnitude of sound gradually weakens, the signal of sound becomes smaller and eventually disappears. However, when light gradually weakens, the overall quantity of its signal becomes less yet the few remaining pulses (extremely short signals) can be detected and the size of these individual signals does not decrease. This tells us that light cannot become any smaller, and that light has a property of a "particle."

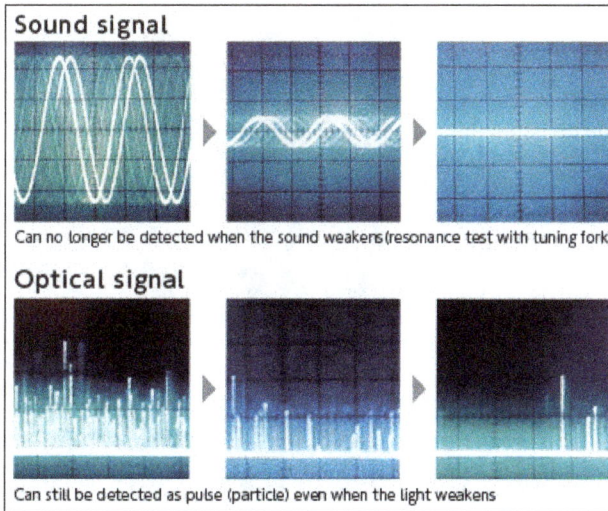

Sound signal

Can no longer be detected when the sound weakens (resonance test with tuning fork)

Optical signal

Can still be detected as pulse (particle) even when the light weakens

Sound signal and Optical signal.

Light Travels at a Speed of 300,000 Kilometers Per Second

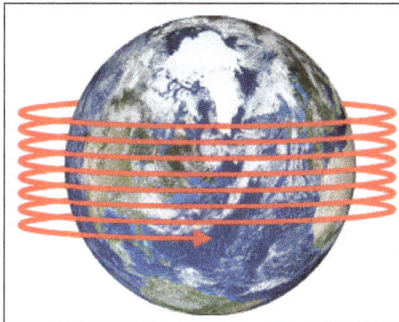

Light travels around the earth 7 and a half times per second.

Light travels at a speed of about 300,000 kilometers per second. Surprisingly, light can travel around the earth 7.5 times in a mere one second. This property of light is utilized in many technical applications such as optical communications which transfer huge data in a very short time. However, even light, which is faster than anything known to man, can move only 0.3 millimeters in a trillionth of a second (a picosecond) in a vacuum. In recent years, research of such optical phenomenon that occur in these unbelievably short period of time, is becoming essential in new research fields of physics, chemistry, biology, and others.

Where, 1 millisecond = 1 thousandth of a second, 1 microsecond = 1 millionth of a second, 1 nanosecond = 1 billionth of a second, 1 picosecond = 1 trillionth of a second.

Interaction of Light with Matter

Light behaves in many different ways when it comes in contact with something.

When in a vacuum such as outer space where no matter is present, light travels straightforward. However, light behaves in a variety of ways when it comes in contact with water, air, and other matters – it is "absorbed", "transmitted through", "reflected", and "scattered". When light strikes matter, a part of that light is absorbed into the matter (a) and is transformed into heat energy. If the matter that the light strikes is a transparent material, the light component that was not absorbed within the material is "transmitted" through (b) and exits to the outer side of the material. If the surface of the material is smooth (a mirror for example), "reflection" occurs (b), but if the surface is irregular having pits and protrusions, the light "scatters" (c).

The "transmitted," "reflected," or "scattered" light allows our eyes to see the colors and shapes of objects.

    a. Absorption:

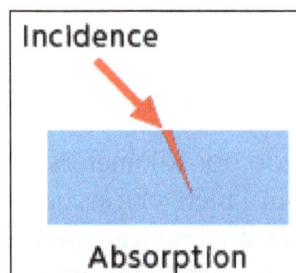

b. Reflection, Transmission:

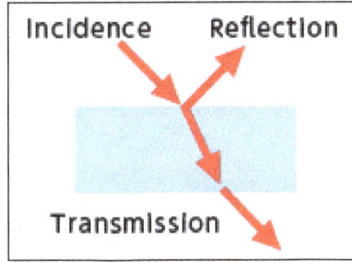

c. Scattering:

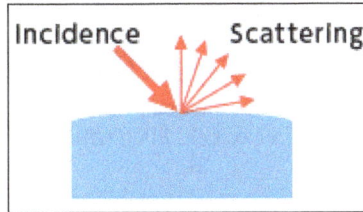

Light "Reflects"

The sunlight striking a mountain bounces back in many directions. This is called reflected light. Our eyes see the mountain by capturing some of the light reflected from the mountain which directly reaches our eyes and then by forming an image of the reflected light on the retina through the lens of the eye. (Pink lines in the figure below represent the reflected light. To make it easier to describe, this figure shows a boy looking at a distant tree instead of a mountain).

When there is a lake or pond between our eyes and a mountain, the light arriving there from the mountain reflects off the surface of the lake or pond (blue dotted lines in the figure). If the surface is calm with no wind and also flat and smooth such as on level surfaces with no irregularities like mirrors and glass, then the angle of the incident light (angle of incidence) and the angle of the light bouncing off the surface (angle of reflection) are equal to each other. This is referred to as specular reflection or mirror reflection. When the surface is located in an ideal location where the light bouncing off the surface by means of specular reflection directly reaches our eyes, then we can see a sharp, clear image of the mountain reflected on the surface.

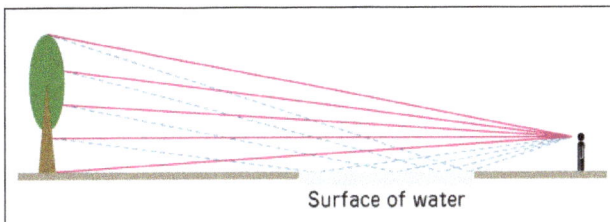

How a landscape appears on the surface of water.

On the other hand, if the surface is rough or irregular, then the direction of the reflected light varies depending on the position on the surface, resulting in a distorted image of the mountain reflecting on the water surface.

A distant mountain is seen reflected on the
surface of a rice field filled with water.

Light "Scatters"

Light from the sun reaches the earth after traveling through space, it "scatters" when striking the various particles and molecules in the atmosphere. A part of this light returns to the outer space and the remainder of the light reaches the surface of the earth after traveling through the atmosphere. The level of scattering of light depends on its wavelength, and of the lights that our eyes can see, blue light is more intensely dispersed or scattered. This is why the sky appears blue to our eyes during the day.

On the other hand, during sunrise and sunset, the sky can appear orange, pink, or red to our eyes. This is because when the position of the sun is lower, the distance that the light travels through the atmosphere becomes longer, and the blue light that is gradually scattered and weakens. Therefore, the remaining red or orange light reaches our eyes.

Blue sky   Sky at sunset

Light "Refracts"

Light "refracts" at the boundary between air and water in the glass. Refraction occurs because light travels at different speed in air and water. Our eyes catch the scattered light from the straw in the water, but refraction occurs when the light in the water enters the air. However, the light coming out from water appears to be moving straightforward to our eyes, and our eyes form a "virtual image" on the line extending from the refracted light. Thus the tip of the straw in the water appears to have deviated from its actual position.

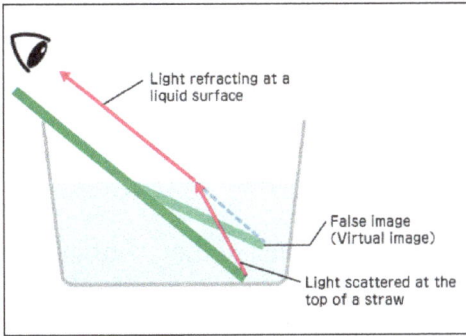

Mechanism that makes a straw in
the water appear bent.

Straw in the glass.

Light "Interferes"

Light moves in various directions so the light waves are constantly striking against each another. The phenomenon that occurs when the light waves collide with each other is called "interference."

When the peaks of these waves overlap, the peaks become even larger. When the peaks and valleys of the waves collide, the waves cancel each other out. This interference is what causes us to see the various colors in soap bubbles. A soap bubble is made of an extremely thin film. Light reflecting from the outer and inner sides of this film interferes with each other to cause the colors that we see. Moreover, the viewing angle of the light interference occurring at the soap bubble film changes due to the ceaseless movement of the soup bubble.

Due to the waves of light repeatedly intensifying and canceling each other out, our eyes see mysterious and constantly changing colors.

Mysterious colors of blowing soap bubbles.

Light "Disperses"

The light from the sun is called white light beam, but it actually is a mixture of different colored lights which appear white to our eyes. Using a prism to separate the white light beam allows us to see the various colors of light.

This phenomenon is called "dispersion" of light. In the natural world, water droplets act like a prism then they remain in the air after the rain. Light that strikes water droplets refracts and moves to the interior of the droplet, reflects within the droplet, and refracts when exiting the droplet. The water droplets in the air act just like a prism causing dispersion and the light reaching our eyes appears as continuous bands of different colors. That is what makes a rainbow.

If we look closely around the rainbow, we may sometimes see another rainbow (a secondary rainbow) whose color sequence is reversed, on the outer side of the first rainbow. This secondary rainbow appears due to light that reaches our eyes reflecting twice in the water droplet.

Rainbow in the sky after the rain.

Laser

The word laser is an acronym for Light Amplification by Stimulated Emission of Radiation. Laser is a device that amplifies or increases the intensity of light and produces highly directional light.

Laser not only amplifies or increases the intensity of light but also generates the light. Laser emits light through a process called stimulated emission of radiation which amplifies or increases the intensity of light. Some lasers generate visible light but others generate ultraviolet or infrared rays which are invisible.

In general, when electron jumps from a higher energy level to a lower energy level, it emits light or photon. The energy of the emitted photon is equal to the energy difference between the energy levels. The loss of electron energy is attributed to the entire atom. Therefore, it can be thought that the atom is moving from a higher energy state to a lower energy state.

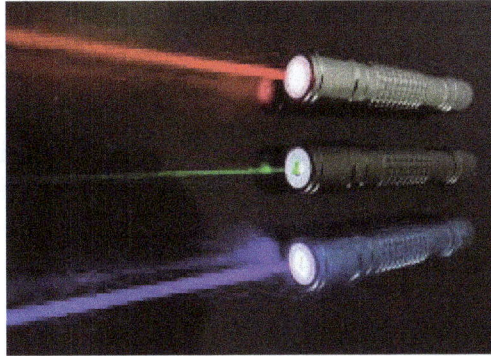

Laser light is different from the conventional light. Laser light has extra-ordinary properties which are not present in the ordinary light sources like sun and incandescent lamp.

The conventional light sources such as electric bulb or tube light does not emit highly directional and coherent light whereas lasers produce highly directional, monochromatic, coherent and polarized light beam. In conventional light sources, excited electrons emit light at different times and in different directions so there is no phase relation between the emitted photons.

On the other hand, the photons emitted by the electrons of laser are in same phase and move in the same direction. Einstein gave the theoretical basis for the development of laser in 1917, when he predicted the possibility of stimulated emission. In 1954, C.H. Townes and his co-workers put Einstein's prediction for practical realization.

They developed a microwave amplifier based on stimulated emission of radiation. It was called as MASER (Microwave Amplification by Stimulated Emission of Radiation.) Maser operates on principles similar to laser but generates microwaves rather than light radiation.

Types of Laser

There are many types of lasers available for different purposes. Depending upon the sources they can be described as below:

1. Solid State Laser: In this kind of lasers solid state, materials are used as active medium. The solid state materials can be ruby, neodymium-YAG (yttrium aluminum garnet) etc.

2. Gas Laser: These lasers contain a mixture of helium and Neon. This mixture is packed up into a glass tube and acts as active medium. We can use Argon or Krypton or Xenon as the medium. CO_2 and Nitrogen Laser can also be made.

3. Dye or Liquid Laser: In this kind of lasers organic dyes like Rhodamine 6G in liquid solution or suspension used as active medium inside the glass tube.

4. Excimer Laser: Excimer lasers (the name came from excited and dimers) use reactive gases like Chlorine and fluorine mixed with inert gases like Argon or Krypton or Xenon. These lasers produce light in the ultraviolet range.

5. Chemical Laser: A chemical laser is a Laser that obtains its energy from a chemical reaction. Examples of chemical lasers are the chemical oxygen iodine laser (COIL), all gas-phase iodine laser (AGIL), and the hydrogen fluoride laser, deuterium fluoride laser etc.

6. Semiconductor Laser: In these Lasers, junction diodes are used. The Semiconductor is doped by both the acceptors and donors. These are known as injection laser diodes. Whenever the current is passed, light can be seen at the output.

Components of Laser

There are four essential or basic components of laser, may be listed as:

1. Active medium,

2. Excitation mechanism,

3. Feedback mechanism,

4. Output coupler.

Active Medium

The active medium is the collection of atoms, ions or molecules in which the stimulated emission occurs. It can be either solid, liquid, gas or semiconductor material. For example, the Ruby laser has a crystal of ruby for its active medium while the CO_2 laser has carbon dioxide gas as active medium.

The wave length emitted by a laser is a function of active medium because the atoms within the active medium have their own characteristics energy level at which they release photons.

Technically speaking, the active medium is the substance that actually laser. For example in He-Ne laser only neon gas laser. Within the laser active medium is usually in the shape of a cylinder or is held within cylindrical container. However other geometries may also be used.

Excitation Mechanisms

The excitation mechanism is a device used to impart or put energy into the active medium.

Now the process of imparting energy to the active medium is called "Pumping the Laser or Energy Pumpling". The energy pumping can be done by three types of excitation mechanism i.e:

1. Optical excitation,

2. Electrical excitation,

3. Chemical excitation.

All these three mechanisms provide the necessary energy to raise the energy state of an atom, ion or molecule of active medium to an excited state.

Optical Excitation

An optical excitation mechanism uses light energy (not electrical energy) of the proper wavelength to excite the active medium. That is why it is called O.E mechanism. Here the light may come from several sources such as flash lamp, continuous arc lam, another laser or even a sun.

Optical excitation is generally used with active medium that do not conduct electricity. So this type of excitation is used exclusively with solid lasers, for example Ruby laser.

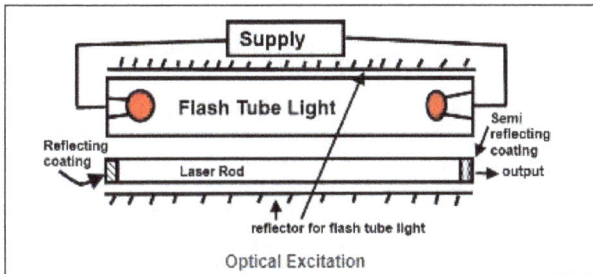

Optical Excitation: Schematic diagram of an optically pumped laser.

Electrical Excitation

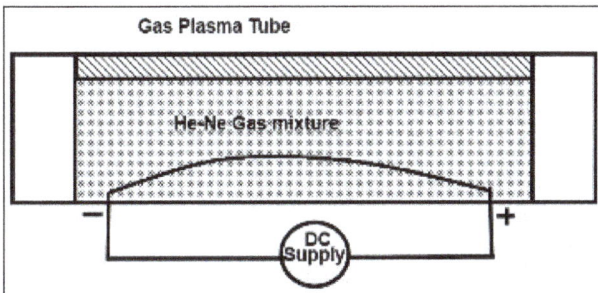

Electrical Excitation: Schematic diagram of electrically excited laser.

An electrical excitation mechanism uses electrical energy to excite the active medium. Best electrical energy source is battery. The electrical excitation is most commonly used when the active medium support an electric current. So this type of excitation is used usually with gas or semiconductor laser. For example He-Ne laser.

When a high voltage is applied to a gas then current carrying electrons or ions move through the active medium carrying energy with them. As they collide with the atoms, ions or molecules of the active medium, then their energy is transferred and excitation occurs. The atoms, ions and electrons within the active medium are called Plasma.

Chemical Excitation

The chemical excitation mechanism uses chemical energy to excite the active medium. For source of chemical energy certain chemicals are mixed, so due to chemical bond formation of broken, energy released. This energy can be used as a pumping source.

The chemical excitation is used in a limited number of lasers for example in hydrogen fluoride lasers, which is extremely high powered device and only used for military applications.

Feedback Mechanisms

The mirrors are used at each end of the active medium as a feedback mechanism. These mirrors reflect the light produced in the active medium back into the medium along its longitudinal axis. When these mirrors are aligned parallel to each other, they form the resonant cavity for the light waves produced within the laser. That is they reflect the light waves back and forth through the active medium.

We know that light is amplified through stimulated emission. So in order to keep stimulated emission at maximum we must keep the light within the amplifying medium for the greatest possible distance. In fact the mirrors increases the distance traveled by the light through the active medium. The path that the light takes through the active medium is determined by the shape of mirrors, as shown by some possible mirror combination.

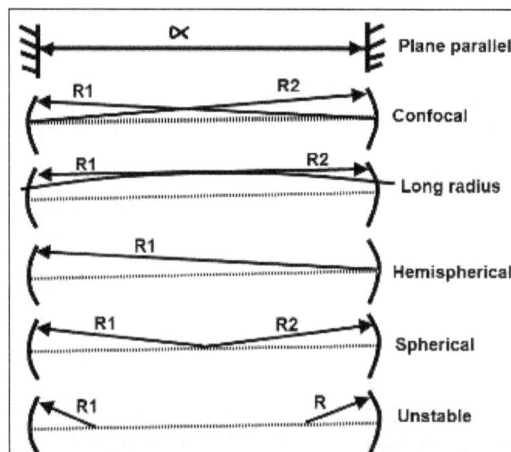

Laser Feedback Mechanism.

From the figure it is clear that both types of mirrors (plane and curved) are used for feedback mechanism. The curved mirrors effects the direction in which the reflected light moves.

Reflectivity from mirror is an important characteristic of laser mirrors. A mirror's reflectivity is its ability to reflect incident light. Mirrors can be designed to reflect just about any percentage of this light. This is necessary because there are a great number of laser types available and they require different mirrors. For instance, in very low power lasers, it sometimes necessary to reflect as much as 99.99% of the laser light back into the resonant cavity in order to keep power losses at a minimum. On other hand, in high power lasers the mirrors reflectivity can be significantly less.

Remember that curvature of mirrors surface is also important. For simplicity and accuracy, the flat mirrors are used in great many lasers, especially high power solid lasers. However it is very common to use curved mirror surfaces. The type and amount of curves must be matched with the distance between the mirrors for best performance.

Output Coupler

The output mirror that is designed to transmit a given percentage of the laser light in the cavity between the feedback mirrors is called the output coupler.

Since the feedback mechanism keeps all the light inside the laser cavity. Now in order to produce the output beam, a portion of the light in the cavity must be allowed to escape by the help of output coupler. This escape must commonly controlled by using a partially reflective mirror in the feedback mechanism. The amount of reflectance required varies with the type of laser. High power lasers may use as little as 35% reflectance with the remaining 65% being transmitted through the mirror (output coupler) to become the output laser beam. A low power laser may require an output mirror reflectivity as high as 98% leaving only 2% to be emitted.

Gain Medium

Within the context of laser physics, a laser gain medium is a medium which can amplify the power of light (typically in the form of a light beam). Such a gain medium is required in a laser to compensate for the resonator losses, and is also called an active laser medium. It can also be used for application in an optical amplifier. The term gain refers to the amount of amplification.

As the gain medium adds energy to the amplified light, it must itself receive some energy through a process called pumping, which may typically involve electrical currents (electrical pumping) or some light inputs ($\rightarrow$ optical pumping), typically at a wavelength which is shorter than the signal wavelength.

Types of Laser Gain Media

There are a variety of very different gain media; the most common of them are:

- Certain direct band gap semiconductors such as gallium arsenide, indium gallium arsenide or gallium nitride are typically pumped with electrical currents, often in the form of quantum wells (→ *semiconductor lasers*).

- Certain laser crystals and glasses such as Nd:YAG (neodymium-doped yttrium aluminum garnet → YAG lasers), Yb:YAG (ytterbium-doped YAG), Yb:-glass, Er:YAG (erbium-doped YAG), or Ti:sapphire are used in the form of solid pieces (→ bulk lasers) or optical glass fibers (→ fiber lasers, fiber amplifiers). These crystals or glasses are doped with some laser-active ions (in most cases trivalent rare earth ions, sometimes transition metal ions) and optically pumped. Lasers based on such media are sometimes called doped insulator lasers.

- There are ceramic gain media, which are also normally doped with rare earth ions.

- Laser dyes are used in dye lasers, typically in the form of liquid solutions.

- Gas lasers are based on certain gases or gas mixtures typically pumped with electrical discharges (e.g. in CO_2 lasers and excimer lasers).

- More exotic gain media are chemical gain media (converting chemical energy to optical energy), nuclear pumped media, and undulators in free electron lasers (transferring energy from a fast electron beam to a light beam).

Compared with most crystalline materials, ion-doped glasses usually exhibit much broader amplification bandwidths, allowing for large wavelength tuning ranges and the generation of ultrashort pulses. Drawbacks are inferior thermal properties (limiting the achievable output powers) and lower laser cross sections, leading to a higher threshold pump power and (for passively mode-locked lasers) to a stronger tendency for Q-switching instabilities.

The doping concentration of crystals, ceramics and glasses often has to be carefully optimized. A high doping density may be desirable for good pump absorption in a short length, but may lead to energy losses related to quenching processes, e.g. caused by upconversion via clustering of laser-active ions and energy transport to defects.

Important Physical Effects

In most cases, the physical origin of the amplification process is stimulated emission, where photons of the incoming beam trigger the emission of additional photons in a process where e.g. initially excited laser ions enter a state with lower energy. Here, there is a distinction between four-level and three-level gain media.

A less frequently used amplification process is stimulated Raman scattering, involving the conversion of some higher-energy pump photons into lower-energy laser photons and phonons (related to vibrations e.g. of the crystal lattice).

For high levels of input light powers, the gain of a gain medium saturates, i.e., is reduced. This naturally follows from the fact that for a finite pump power an amplifier cannot add arbitrary amounts of power to an input beam. In laser amplifiers, saturation is related to a decrease in population in the upper laser level, caused by stimulated emission.

Thermal effects can occur in gain media, because part of the pump power is converted into heat. The resulting temperature gradients and also subsequent mechanical stress can cause lensing effects, distorting the amplified beam. Such effects can spoil the beam quality of a laser, reduce its efficiency, and sometimes even destroy the gain medium (thermal fracture).

Relevant Physical Properties of Laser Gain Media

A great variety of physical properties of a gain medium can be relevant for use in a laser. The desirable properties include:

- A laser transition in the desired wavelength region, preferably with the maximum gain occurring in this region.

- A high transparency of the host medium in this wavelength region.

- A pump wavelength for which a good pump source is available (in case of an optically pumped laser); efficient pump absorption.

- A suitable upper-state lifetime: Long enough for q-switching applications, short enough if fast modulation of the power is required.

- A high quantum efficiency, obtained via a low prevalence for quenching effects, excited-state absorption and the like, but also possibly by strong enough beneficial effects such as certain multi-phonon transitions or energy transfers.

- Ideally, four-level behavior, because quasi-three-level behavior introduces various additional constraints.

- Robustness and a long lifetime, chemical stability.

- For solid-state gain media: A host medium which is available with good optical quality in the required size, can be cut and polished with good quality (appropriate hardness), allows for high doping with laser-active ions without clustering, is chemically stable (e.g., not hygroscopic), and has a good thermal conductivity and low thermo-optic coefficients (for weak thermal lensing in high-power operation) and high resistance to mechanical stress; optical isotropy can be desirable, but in other cases birefringence (reducing thermal depolarization) and possibly polarization-dependent gain is preferable.

- For high gain, low threshold pump power: A high product of emission cross section and upper-state lifetime (σ–τ product).

- For low beam quality requirements on the pump source: High pump absorption may be helpful.

- For wavelength tuning: A large gain bandwidth.

- For ultrashort pulse generation: A broad and smoothly shaped gain spectrum; suitable chromatic dispersion and nonlinearity are also sometimes important.

- For passive mode locking without q-switching instabilities: High enough laser cross sections.

- For high energy pulse amplification (e.g. in regenerative amplifiers): A high optical damage threshold and not too high saturation fluence of the gain.

In many situations there are partially conflicting requirements. For example, a very low quantum defect is not compatible with four-level behavior. A large gain bandwidth typically means that laser cross sections are smaller than ideal, and that the quantum defect cannot be very small. Disorder in solid-state gain media increases the gain bandwidth, but also reduces the thermal conductivity. A short pump absorption length can be advantageous, but also tends to exacerbate thermal effects.

It is apparent that different situations lead to very different requirements on gain media. For this reason, a very broad range of gain media will continue to remain important for applications, and making the right choice is essential for constructing lasers with optimum performance.

Population Inversion

Population inversion is the process of achieving greater population of higher energy state as compared to the lower energy state. Population inversion technique is mainly used for light amplification. The population inversion is required for laser operation.

Consider a group of electrons with two energy levels E_1 and E_2:

- E_1 is the lower energy state and E_2 is the higher energy state.

- N_1 is the number of electrons in the energy state E_1.

- N_2 is the number of electrons in the energy state E_2.

The number of electrons per unit volume in an energy state is the population of that energy state.

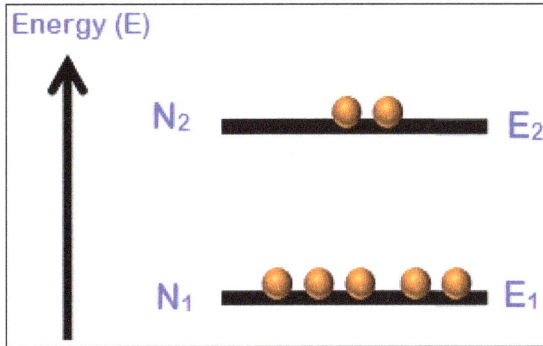

Population inversion cannot be achieved in a two energy level system. Under normal conditions, the number of electrons (N_1) in the lower energy state (E_1) is always greater as compared to the number of electrons (N_2) in the higher energy state (E_2).

$N_1 > N_2$

When temperature increases, the population of higher energy state (N_2) also increases. However, the population of higher energy state (N_2) will never exceeds the population of lower energy state (N_1).

At best an equal population of the two states can be achieved which results in no optical gain.

$N_1 = N_2$

Therefore, we need 3 or more energy states to achieve population inversion. The greater is the number of energy states the greater is the optical gain.

There are certain substances in which the electrons once excited; they remain in the higher energy level or excited state for longer period. Such systems are called active systems or active media which are generally mixture of different elements.

When such mixtures are formed, their electronic energy levels are modified and some of them acquire special properties. Such types of materials are used to form 3-level laser or 4-level laser.

3-level Laser

Consider a system consisting of three energy levels E_1, E_2, E_3 with N number of electrons.

We assume that the energy level of E_1 is less than than E_2 and E_3, the energy level of E_2 is greater than E_1 and less than E_3, and the energy level of E_3 is greater than E_1 and E_2.

It can be simply written as $E_1 < E_2 < E_3$. That means the energy level of E_2 lies in between E_1 and E_3.

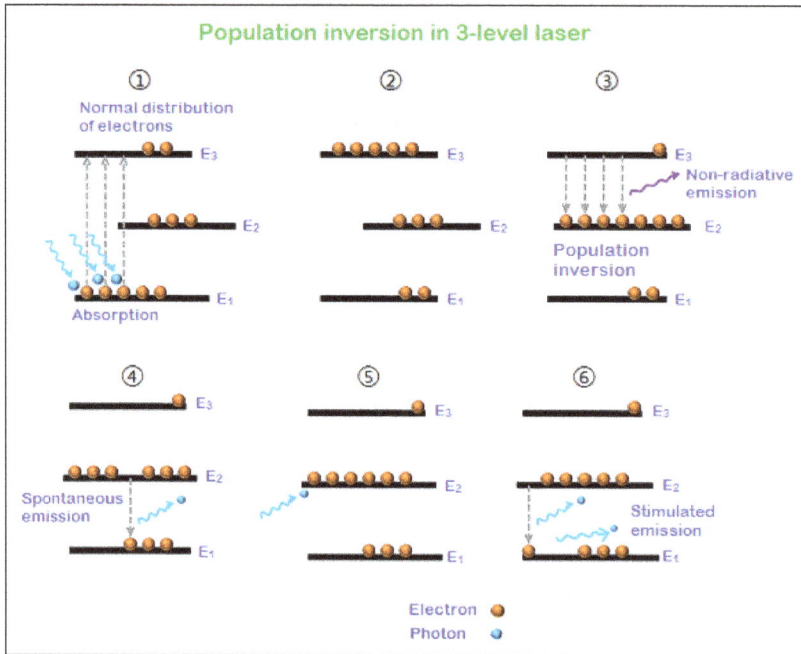

Population inversion in 3-level laser.

The energy level E_1 is known as the ground state or lower energy state and the energy levels E_2 and E_3 are known as excited states. The energy level E_2 is sometimes referred to as Meta stable state. The energy level E_3 is sometimes referred to as pump state or pump level.

The N number of electrons in the system occupies these three energy levels. Let N_1 be the number of electrons in the energy state E_1, N_2 be the number of electrons in the energy state E_2 and N_3 be the number of electrons in the energy state E_3. To get laser emission or population inversion, the population of higher energy state (E_2) should be greater than the population of the lower energy state (E_1).

Under normal conditions, the higher an energy level is, the lesser it is populated. For example, in a three level energy system, the lower energy state E_1 is highly populated as compared to the excited energy states E_2 and E_3. On the other hand, the excited energy state E_2 is highly populated as compared to the excited energy state E_3. It can be simply written as $N_1 > N_2 > N_3$. Under certain conditions, the greater population of higher energy state (E_2) as compared to the lower energy state (E_1) is achieved. Such an arrangement is called population inversion.

Let us assume that initially the majority of electrons will be in the lower energy state or ground state (E_1) and only a small number of electrons will be in excited states (E_2 and E_3).

When we supply light energy which is equal to the energy difference of E_3 and E_1, the electrons in the lower energy state (E_1) gains sufficient energy and jumps into the higher energy state (E_3). This process of supplying energy is called pumping. We also use other methods to excite ground state electrons such as electric discharge and chemical reactions. The flow of electrons from E_1 to E_3 is called pump transition.

The lifetime of electrons in the energy state E_3 is very small as compared to the lifetime of electrons in the energy state E_2. Therefore, electrons in the energy level E_3 does not stay for long period. After a short period, they quickly fall to the Meta stable state or energy state E_2 and releases radiation less energy instead of photons. Because of the shorter lifetime, only a small number of electrons accumulate in the energy state E_3.

The electrons in the Meta stable state E_2 will remain there for longer period because of its longer lifetime. As result, a large number of electrons accumulate in Meta stable state. Thus, the population of metal stable state will become greater than the population of energy states E_3 and E_1. It can be simply written as $N_2 > N_1 > N_3$.

In a three level energy system, we achieve population inversion between energy levels E_1 and E_2.

After completion of lifetime of electrons in the Meta stable state, they fall back to the lower energy state or ground state E_1 by releasing energy in the form of photons. This process of emission of photons is called spontaneous emission.

When this emitted photon interacts with the electron in the Meta stable state E_2, it forces that electron to fall back to the ground state. As a result, two photons are emitted. This process of emission of photons is called stimulated emission. When these photons again interacted with the electrons in the Meta stable state, they forces two Meta stable state electrons to fall back to the ground state. As a result, four photons are emitted. Likewise, a large number of photons are emitted.

As a result, millions of photons are emitted by using small number of photons. We may get a doubt, in order to excite an electron we hit the electron with a photon. This excited electron again emits photon when fall back to the ground state. Then how could light amplification or extra photons is achieved. We may also use other types of energy sources such as electrical energy to excite electrons. In such case, a single photon will generates large number of photons. Thus, light amplification is achieved by using population inversion method. The system which uses three energy levels is known as 3-level laser.

In a 3-level laser, at least half the population of electrons must be excited to the higher energy state to achieve population inversion. Therefore, the laser medium must be very strongly pumped. This makes 3-level lasers inefficient to produce photons or light. The three level lasers are the first type of lasers discovered.

4-level Laser

Consider a group of electrons with four energy levels E_1, E_2, E_3, E_4. E_1 is the lowest energy state, E_2 is the next higher energy, E_3 is the next higher energy state after E_2, E_4 is the next higher energy state after E_3.

The number of electrons in the lower energy state or ground state is given by N_1, the number of electrons in the energy state E_2 is given by N_2, the number of electrons in the energy state E_3 is given by N_3 and the number of electrons in the energy state E_4 is given by N_4. We assume that $E_1 < E_2 < E_3 < E_4$. The lifetime of electrons in the energy state E_4 and energy state E_2 is very less. Therefore, electrons in these states will only stay for very short period.

When we supply light energy which is equal to the energy difference of E_4 and E_1, the electrons in the lower energy state E_1 gains sufficient energy and jumps into the higher energy state E_4. The lifetime of electrons in the energy state E_4 is very small. Therefore, after a short period they fall back into the next lower energy state E_3 by releasing non-radiation energy.

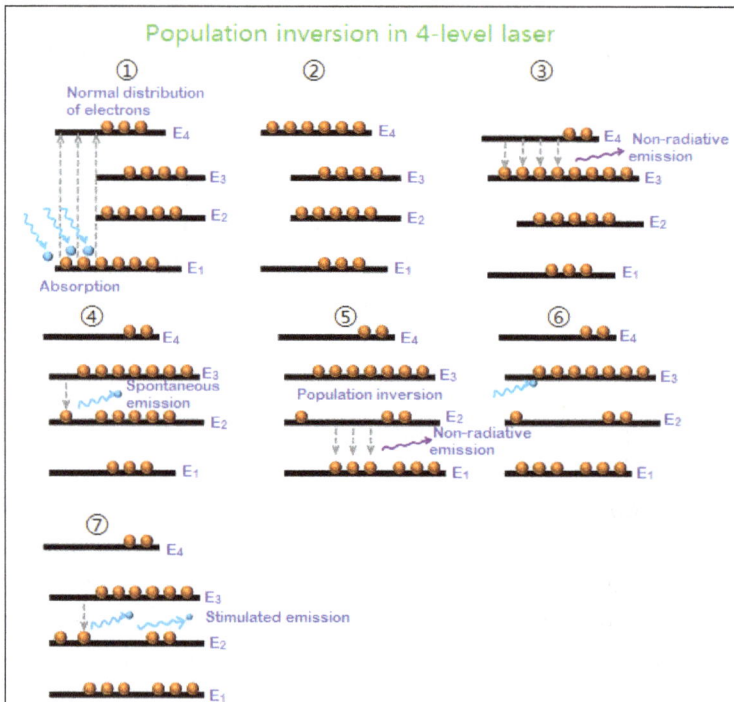

Population inversion in 4-level laser.

The lifetime of electrons in the energy state E_3 is very large as compared to E_4 and E_2. As a result, a large number of electrons accumulate in the energy level E_3. After completion of their lifetime, the electrons in the energy state E_3 will fall back into the next lower energy state E_2 by releasing energy in the form of photons.

Like the energy state E_4, the lifetime of electrons in the energy state E_2 is also very small. Therefore, the electrons in the energy state E_2 will quickly fall into the next lower energy state or ground state E_1 by releasing non-radiation energy.

Thus, population inversion is achieved between energy states E_3 and E_2.

In a 4-level laser, only a few electrons are excited to achieve population inversion. Therefore, a 4-level laser produces light efficiently than a 3-level laser. In practical, more than four energy levels may be involved in the laser process. In 3-level and 4-level lasers, the frequency or energy of the pumping photons must be greater than the emitted photons.

Saturation

The saturation power of a laser gain medium is the optical power of an input signal which *in the steady state* leads to a reduction in the gain to half of its small-signal value. Similarly, the saturation power of a saturable absorber is defined. The saturation intensity is the corresponding optical intensity, i.e., the saturation power per unit area.

Usually it is assumed that the gain is small, i.e. input and output powers are similar. For high gain, it is common to refer to the output power.

For a low-gain laser amplifier, saturation intensity and power can be calculated according to:

$$l_{sat} = \frac{h\nu}{\left(\sigma_{em} + \sigma_{abs}\right)\tau}, P_{sat} = Al_{sat} = \frac{Ah\nu}{\left(\sigma_{em} + \sigma_{abs}\right)\tau}$$

where $h\nu$ is the photon energy at the signal wavelength, σ_{em} and σ_{abs} are the emission and absorption cross sections at the emission wavelength, τ is the upper-state lifetime, and A is the mode area. The quantity σ_{abs} is zero for four-level gain media but should not be forgotten for quasi-three-level gain media.

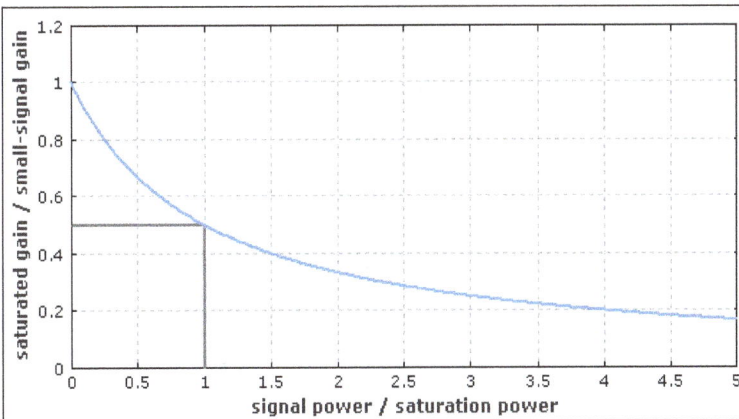

Dependence of laser gain on the optical power, calculated
for the steady state. When the power equals the saturation power,
the gain is reduced to half the small-signal gain.

A comparison with the equations for the saturation energy shows that the saturation power can be calculated as the saturation energy divided by the upper-state lifetime.

Importance of the Saturation Power

The saturation power plays an important role in various areas of laser physics and laser or amplifier design. Some examples are:

- It determines the amount of input power of an amplifier required for achieving most of the possible output power.

- The laser intensity in the gain medium of a four-level laser equals the saturation intensity if the pump power is twice the threshold pump power. This is remarkable, because the laser intensity in this situation is thus determined only by a property of the gain medium, not by resonator losses etc.

- For a saturable absorber, as used e.g. in a mode-locked laser, the ratio of continuous-wave intracavity power to saturation power is an important parameter for the initial pulse formation process.

The saturation power should not be confused with the *saturated output power*, which is usually not precisely defined but means the output power achieved for an input signal power which causes significant amplifier saturation. Obviously, the saturated output power (other than the saturation power) depends on the pump power.

The Rate Equation Modeling

The dynamics of energy level populations in laser gain media (e.g. rare-earth-doped crystals or fibers) are often modeled using a system of *rate equations*. These are differential equations, describing the temporal evolution of level populations under the influence of optically induced and non-radiative transitions:

- Absorption processes, possibly including excited-state absorption.

- Spontaneous and stimulated emission.

- Multi-phonon transitions.

- Energy transfers which can lead to upconversion and quenching.

Rate equation models can be part of more comprehensive numerical models, which describe e.g. the spatial distribution of optical powers in fiber amplifiers or bulk lasers, or the dynamic behavior of Q-switched lasers. They can thus help in understanding quantitatively the operation of laser and amplifier devices, and allow one, e.g., to evaluate whether the performance of a device is close to the limitations set by the fundamental principle of operation.

Example: Rate Equations for Erbium-doped Gain Media.

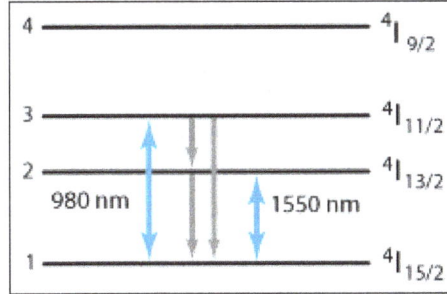

Energy levels and transitions in an erbium-doped gain medium.

As an example, consider the dynamics of an erbium-doped gain medium, such as used in, e.g., erbium-doped fiber amplifiers. Figure shows the energy level diagram and the most important radiative and non-radiative transitions. Erbium amplifiers and lasers operating in the 1.5-μm spectral region exhibit a quasi-three-level scheme. For simplicity, the energy levels (actually Stark level manifolds) of the erbium (Er^{3+}) ions are labeled on the left-hand side with an index which is e.g. for the ground-state manifold $^4I_{15/2}$ and 2 for the upper-state manifold $^4I_{13/2}$. Optical absorption and stimulated emission transitions, as caused by a pump beam at 980 nm and a signal beam at 1550 nm, are indicated with blue arrows, and the gray arrows indicate spontaneous and non-radiative transitions. The lower transition (level 2 → 1) is mostly caused by spontaneous emission, whereas the upper transition (3 → 2) is dominated by multi-phonon transitions. Only transitions between the lowest three levels are considered, assuming that excited-state absorption to higher levels and upconversion processes are weak.

For that situation, the rate equation system obtained is the following:

$$\frac{\partial n_3}{\partial t} = -A_{32}n_3 - A_{31}n_3 + \frac{\sigma_{13}I_p}{hv_p}n_1 - \frac{\sigma_{31}I_p}{hv_p}n_3$$

$$\frac{\partial n_2}{\partial t} = -A_{21}n_2 + A_{32}n_3 + \frac{\sigma_{12}I_s}{hv_s}n_1 - \frac{\sigma_{21}I_s}{hv_s}n_2$$

$$\frac{\partial n_1}{\partial t} = +A_{21}n_2 + A_{31}n_3 - \frac{\sigma_{13}I_p}{hv_p}n_1 + \frac{\sigma_{31}I_p}{hv_p}n_3 - \frac{\sigma_{12}I_s}{hv_s}n_1 + \frac{\sigma_{21}I_s}{hv_s}n_2$$

where n_j indicates the fractional level population of level j. This variable is e.g. if all ions are in the corresponding level manifold. As only the mentioned three levels are involved, we have $n_1 + n_2 + n_3 = 1$. (The parameters n_j can also be interpreted as excitation densities with units of m^{-3}; only in rate equation systems containing nonlinear terms, e.g. for energy transfers, the form of the equations depends on that interpretation.) The parameters A_{jk} indicate spontaneous transition rates from level j to k, with units of s^{-1}. For example, A_{21} is the inverse upper-state lifetime. Furthermore, the equations contain absorption and stimulated emission rates, which are determined by transition cross sections σ_{jk} (the values of which are dependent on the wavelengths involved), optical intensities I_p and I_s at the pump and signal wavelength, and photon energies $h\nu$.

The multi-phonon transition $3 \to 2$ is usually strong if the gain medium (e.g. a silica fiber) has a high phonon energy. Therefore, n_3 will usually be small (except for extremely high pump intensities), as ions pumped into level 3 will rapidly be transferred to level 2. Consequently, it is often valid to neglect the population in level 3, and also the other transitions starting from that level. The equation system then simplifies to,

$$\frac{\partial n_2}{\partial t} = -A_{21} n_2 + \frac{\sigma_{13} I_p}{h\nu_p} n_1 + \frac{\sigma_{12} I_s}{h\nu_s} n_1 - \frac{\sigma_{21} I_s}{h\nu_s} n_2$$

$$\frac{\partial n_1}{\partial t} = +A_{21} n_2 - \frac{\sigma_{13} I_p}{h\nu_p} n_1 - \frac{\sigma_{12} I_s}{h\nu_s} n_1 + \frac{\sigma_{21} I_s}{h\nu_s} n_2$$

where of course one of the equations is redundant, as the sum of the two level populations must stay unity in any case.

Various circumstances can lead to additional complications:

- It is common to insert additional terms, e.g. for energy transfer processes. There may be, e.g., terms proportional to $n_2{}^2$ for cooperative up conversion processes, where one ion in level 2 transfers energy to another ion in the same level, resulting in one more ion in the ground-state manifold and the other one in a higher level. Assuming that the higher levels quickly decay to level 2, the additions to the equation system are relatively simple.

- It is also possible to include energy transfers between different species of ions. A common case is that of erbium–ytterbium-doped fibers, where primarily ytterbium ions absorb pump radiation, and transfer energy to erbium ions. The model then includes level populations for all involved types of ions, and the corresponding coupling terms.

- Additional optical wavelengths can be involved e.g. if amplified spontaneous emission (ASE) occurs in a fiber amplifier. In numerical models, the ASE spectrum is divided into discrete wavelength slots, each one being associated with different values of the absorption and emission cross sections.

- Generally, the rate equations have to be solved for different positions within the gain medium, as the optical intensities depend on both the longitudinal and transverse coordinates.

Solving and using the Rate Equations

For a given location in the gain medium and for given optical intensities, the temporal evolution of the population densities can be calculated by temporal integration of the rate equations. This can be done e.g. with the Runge–Kutta method. Note, however, that rate equations are often a so-called "stiff" system of differential equations, involving very different time constants. The fastest processes in the system (e.g. related to relatively fast non-radiative transitions) then force one to use a fairly small temporal step size, as the numerical solution would otherwise become unstable. At the same time, much slower processes often imply that the evolution must be calculated over a long time span, so that in effect many numerical steps are needed. For such reasons, it can be inefficient to calculate the steady-state population densities for given optical intensities simply by simulating the temporal evolution.

For simple cases, it is easy to calculate analytically the steady-state populations for given pump and signal intensities. For example, the simplified equation system above leads to the following result:

$$n_2 = \frac{\dfrac{\sigma_{13}I_p}{h\nu_p} + \dfrac{\sigma_{12}I_s}{h\nu_s}}{A_{21} + \dfrac{\sigma_{13}I_p}{h\nu_p} + \dfrac{(\sigma_{12} + \sigma_{21})I_s}{h\nu_s}}$$

It is also possible to calculate the evolution of both the population densities and the optical powers, e.g., in order to model the laser dynamics of a Q-switched laser. The set of differential equations then includes the rate equations for the populations as well as dynamical equations for the optical powers.

In many situations, it is possible and convenient to derive equations for spatially averaged population densities. In the case of a simple level scheme, the population of the gain medium is then described with a single variable, which is directly related to the laser gain. This variable can then be used e.g. as a dynamic variable in the dynamic equations for population (or gain) and laser intensity.

Limitations of Rate Equation Modeling

Rate equations describe the statistical evolution of level populations, and are based on a number of assumptions. They do not describe coherent phenomena such as Rabi oscillations, as they average over many ions, which experience slightly different microscopic environments and different optical intensities. It is further assumed that all ions essentially function in the same way. The latter assumption can be violated e.g. if clustering occurs in the gain medium. In such a case, ions within clusters may exhibit, e.g., much stronger up conversion processes than other ions do, and should therefore be treated in the model as a separate species. Such extended models have been developed, but they are more complicated, and often involve a number of parameters which are hard to access experimentally.

References

- Light, science: britannica.com, Retrieved 21 July, 2019

- Behavior, photon: photonterrace.net, Retrieved 15 January, 2019

- Laserintroduction, laser, physics: physics-and-radio-electronics.com, Retrieved 3 May, 2019

- Laser-types-and-components-of-laser: electrical4u.com, Retrieved 19 May, 2019

- Essential-laser-components, microwave-radar, electronics: daenotes.com, Retrieved 3 June, 2019

- Gain-media: rp-photonics.com, Retrieved 13 July, 2019

- Laser-populationinversion, laser, physics: physics-and-radio-electronics.com, Retrieved 17 May, 2019

- Rate-equation-modeling: rp-photonics.com, Retrieved 13 February, 2019

2

Types of Lasers

There are numerous types of lasers. Some of them are solid state laser, dye laser, semiconductor laser, gas laser, chemical laser, copper vapor laser, argon ion laser, krypton ion laser and free electron laser. The topics elaborated in this chapter will help in gaining an extensive understanding about the diverse aspects of these types of lasers.

Solid State Laser

Solid-state lasers are the original, conventional workhorses of the bench-top world, differing from semiconductor lasers in size, function, and application.

Solid-state lasers are lasers whose active medium is typically a minority ion in a solid-state host. The host is most often a single crystal with about 1% of a different species, such as a neodymium ion (Nd^{3+}), doped into the solid matrix of the host. For certain applications the host is a glass, the most important of these being glass doped with the erbium ion (Er^{3+}), which is then pulled into a long thin fiber and used in telecommunications systems. Solid-state lasers use optical pumping by other lasers or lamps to produce a wide variety of laser devices. The population inversion (a condition in which the atomic population of the higher energy state is greater than the lower energy state) in solid-state lasers is created when the active medium absorbs photons from an intense light source.

One important class of solid-state lasers consists of the tunable and ultrafast lasers. These systems use solid-state lasers to excite them, and have other similarities in their thermal and optical properties.

The typical optically pumped, solid-state laser system is much larger than a semiconductor device. The optical pump may either be a single, "high-power" semiconductor diode, a small array of semiconductor diodes on one chip, or a stack of arrays for a laser with outputs of up to 1000 W. Before the advent of laser diodes, solid-state lasers were pumped by lamps—either running continuously for modest power lasers, or pulsed flash lamps for laser output energy in the 1-Joule-per-pulse category. Since the number of arrays needed to power a 1-J-per-pulse laser is still quite pricey, lamps still have an important niche in the solid-state laser business.

The active medium of a solid-state laser consists of a passive host crystal and the active ion, and it is these components that give the laser its name. A "neodymium:YAG" (Nd:YAG) laser, for example, consists of a crystal of yttrium aluminum garnet (YAG) with a small amount of neodymium added as an impurity. It is the Nd ion (Nd^{3+} added in the form of Nd_2O_3 to the materials to make the single crystal) in which the population inversion is created, and which generates the photon of laser light. Typically the lasing ion is present at about 0.1% to 1% of the ion density of the metal ions of the host crystal or glass. There are times, however, when the host of the active ion in the solid-state laser is used as the name of the laser. The Nd:YAG is often called a YAG laser, harking back to a time when the only good solid-state laser was YAG with Nd^{3+} ion minority constituent.

Today, however, there are perfectly good solid-state lasers that use YAG as the host for Yb (ytterbium, lasing at 1.03 µm), Ho (holmium, 2.1 µm), Tm (thulium, 2 µm) or Er (erbium, 2.9 µm) ions. In all of these cases the host material is usually configured in a simple rod shape that is cut from a synthetically grown crystal or poured glass. For the important case of the erbium-glass medium, the glass with erbium minority constituent is drawn into a very long thin fiber and typically used as an amplifier in an erbium doped fiber amplifier (EDFA). In general, the vendor of a laser crystal is not the manufacturer of the laser itself.

Ruby Laser

A ruby laser is a solid-state laser that uses the synthetic ruby crystal as its laser medium. Ruby laser is the first successful laser developed by Maiman in 1960.

Ruby laser is one of the few solid-state lasers that produce visible light. It emits deep red light of wavelength 694.3 nm.

Construction of Ruby Laser

A ruby laser consists of three important elements: laser medium, the pump source, and the optical resonator.

Laser Medium or Gain Medium in Ruby Laser

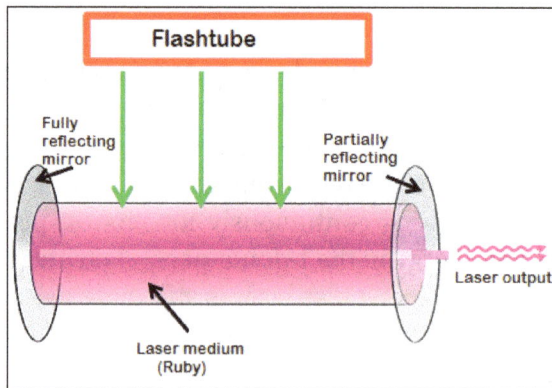

In a ruby laser, a single crystal of ruby (Al_2O_3: Cr^{3+}) in the form of cylinder acts as a laser medium or active medium. The laser medium (ruby) in the ruby laser is made of the host of sapphire (Al_2O_3) which is doped with small amounts of chromium ions (Cr^{3+}). The ruby has good thermal properties.

Pump Source or Energy Source in Ruby Laser

The pump source is the element of a ruby laser system that provides energy to the laser medium. In a ruby laser, population inversion is required to achieve laser emission. Population inversion is the process of achieving the greater population of higher energy state than the lower energy state. In order to achieve population inversion, we need to supply energy to the laser medium (ruby).

In a ruby laser, we use flashtube as the energy source or pump source. The flashtube supplies energy to the laser medium (ruby). When lower energy state electrons in the laser medium gain sufficient energy from the flashtube, they jump into the higher energy state or excited state.

Optical Resonator

The ends of the cylindrical ruby rod are flat and parallel. The cylindrical ruby rod is placed between two mirrors. The optical coating is applied to both the mirrors. The process of depositing thin layers of metals on glass substrates to make mirror surfaces is called silvering. Each mirror is coated or silvered differently.

At one end of the rod, the mirror is fully silvered whereas, at another end, the mirror is partially silvered.

The fully silvered mirror will completely reflect the light whereas the partially silvered mirror will reflect most part of the light but allows a small portion of light through it to produce output laser light.

Working of Ruby Laser

The ruby laser is a three level solid-state laser. In a ruby laser, optical pumping technique is used to supply energy to the laser medium. Optical pumping is a technique in which light is used as energy source to raise electrons from lower energy level to the higher energy level. Consider a ruby laser medium consisting of three energy levels E_1, E_2, E_3 with N number of electrons.

We assume that the energy levels will be $E_1 < E_2 < E_3$. The energy level E_1 is known as ground state or lower energy state, the energy level E_2 is known as metastable state, and the energy level E_3 is known as pump state. Let us assume that initially most of the electrons are in the lower energy state (E_1) and only a tiny number of electrons are in the excited states (E_2 and E_3).

When light energy is supplied to the laser medium (ruby), the electrons in the lower energy state or ground state (E_1) gains enough energy and jumps into the pump state (E_3).

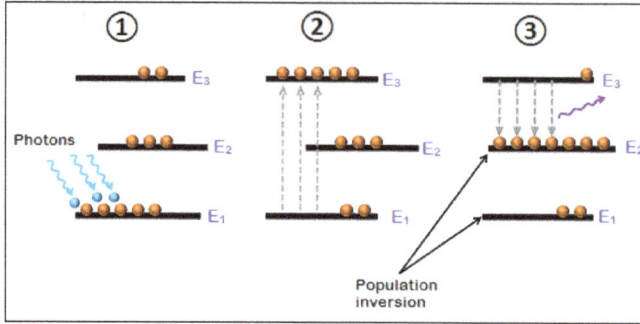

The lifetime of pump state E_3 is very small (10^{-8} sec) so the electrons in the pump state do not stay for long period. After a short period, they fall into the metastable state E_2 by releasing radiationless energy. The lifetime of metastable state E_2 is 10^{-3} sec which is much greater than the lifetime of pump state E_3. Therefore, the electrons reach E_2 much faster than they leave E_2. This results in an increase in the number of electrons in the metastable state E_2 and hence population inversion is achieved.

After some period, the electrons in the metastable state E_2 falls into the lower energy state E_1 by releasing energy in the form of photons. This is called spontaneous emission of radiation.

When the emitted photon interacts with the electron in the metastable state, it force-fully makes that electron fall into the ground state E_1. As a result, two photons are emitted. This is called stimulated emission of radiation. When these emitted photons again interacted with the metastable state electrons, then 4 photons are produced. Because of this continuous interaction with the electrons, millions of photons are produced.

In an active medium (ruby), a process called spontaneous emission produces light. The light produced within the laser medium will bounce back and forth between the two mirrors. This stimulates other electrons to fall into the ground state by releasing light energy. This is called stimulated emission. Likewise, millions of electrons are stimulated to emit light. Thus, the light gain is achieved. The amplified light escapes through the partially reflecting mirror to produce laser light.

Neodymium Doped Yttrium Aluminum Garnet Laser

Neodymium-doped Yttrium Aluminum Garnet (Nd: YAG) laser is a solid state laser in which Nd: YAG is used as a laser medium.

These lasers have many different applications in the medical and scientific field for processes such as Lasik surgery and laser spectroscopy. Nd: YAG laser is a four-level laser system, which means that the four energy levels are involved in laser action. These lasers operate in both pulsed and continuous mode. Nd: YAG laser generates laser light commonly in the near-infrared region of the spectrum at 1064 nanometers (nm). It also emits laser light at several different wavelengths including 1440 nm, 1320 nm, 1120 nm, and 940 nm.

Properties of Nd:YAG

Nd^{3+}:YAG is a four-level gain medium (except for the 946-nm transition as discussed below), offering substantial laser gain even for moderate excitation levels and pump intensities. The gain bandwidth is relatively small, but this allows for a high gain efficiency and thus low threshold pump power.

Nd:YAG lasers can be diode pumped or lamp pumped. Lamp pumping is possible due to the broadband pump absorption mainly in the 800-nm region and the four-level characteristics.

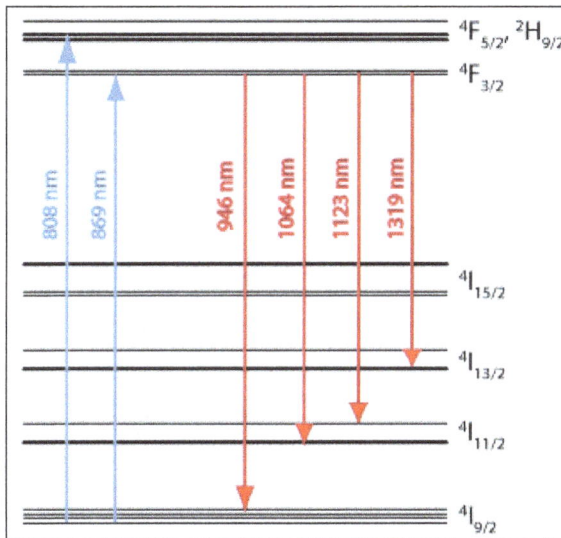

Energy level structure and common pump and laser transitions of the trivalent neodymium ion in Nd^{3+}:YAG.

The most common Nd:YAG emission wavelength is 1064 nm. Starting with that wavelength, outputs at 532, 355 and 266 nm can be generated by frequency doubling, frequency tripling and frequency quadrupling, respectively. Other emission lines are at 946, 1123, 1319, 1338, 1415 and 1444 nm. When used at the 946-nm transition, Nd:YAG is a quasi-three-level gain medium, requiring significantly higher pump intensities. All other transitions are four-level transitions. Some of these, such as the one at 1123 nm, are very weak, so that efficient laser operation on these wavelengths is difficult to obtain:

- Even a moderate gain requires a high excitation density, which favors detrimental quenching effects.

- In addition, lasing at 1064 nm, the wavelength with much higher gain, has to be suppressed, for example by using suitable dichroic mirrors for building the laser resonator.

However, with careful optimization, even on these weak transitions one can obtain substantial output powers.

Nd:YAG is usually used in monocrystalline form, fabricated with the Czochralski growth method, but there is also ceramic (polycrystalline) Nd:YAG available in high quality and in large sizes. For both monocrystalline and ceramic Nd:YAG, absorption and scattering losses within the length of a laser crystal are normally negligible, even for relatively long crystals.

Typical neodymium doping concentrations are of the order of 1 at %. High doping concentrations can be advantageous e.g. because they reduce the pump absorption length, but too high concentrations lead to quenching of the upper-state lifetime e.g. via upconversion processes. Also, the density of dissipated power can become too high in high-power lasers. Note that the neodymium doping density does not necessarily have to be the same in all parts; there are composite laser crystals with doped and undoped parts, or with parts having different doping densities.

Table: Some properties of YAG = yttrium aluminum garnet, which are similar for Nd- or Yb-doped YAG.

Property	Value
Chemical formula	$Y_3Al_5O_{12}$
Crystal structure	cubic
Mass density	4.56 g/cm³
Moh hardness	8–8.5
Young's modulus	280 GPa
Tensile strength	200 MPa
Melting point	1970 °C
Thermal conductivity	10–14 W/(m K)
Thermal expansion coefficient	7–$8 \cdot 10^{-6}$/K
Thermal shock resistance parameter	790 W/m
Birefringence	None (only thermally induced)
Refractive index at 1064 nm	1.82
Temperature dependence of refractive index	7–$10 \cdot 10^{-6}$/k

Table: Some properties of Nd:YAG = neodymium-doped yttrium aluminum garnet.

Property	Value
Nd density for 1 at. % doping	$1.36 \cdot 10^{20}$ cm⁻³
Fluorescence lifetime	230 µs
Absorption cross section at 808 nm	$7.7 \cdot 10^{-20}$ cm²
Emission cross section at 946 nm	$5 \cdot 10^{-20}$ cm²
Emission cross section at 1064 nm	$28 \cdot 10^{-20}$ cm²
Emission cross section at 1319 nm	$9.5 \cdot 10^{-20}$ cm²

Emission cross section at 1338 nm	$10 \cdot 10^{-20}\,cm^2$
Gain bandwidth	0.6 nm

Table: Some properties of Yb:YAG = ytterbium-doped yttrium aluminum garnet.

Property	Value
Yb density for 1 at.% doping	$1.38 \cdot 10^{20}\,cm^{-3}$
Fluorescence lifetime	950 µs
Absorption cross section at 940 nm	$0.75 \cdot 10^{-20}\,cm^2$
Emission cross section at 1030 nm	$2.2 \cdot 10^{-20}\,cm^2$
Absorption cross section at 1030 nm	$0.12 \cdot 10^{-20}\,cm^2$
Emission cross section at 1050 nm	$0.3 \cdot 10^{-20}\,cm^2$
Absorption cross section at 1050 nm	$0.01 \cdot 10^{-20}\,cm^2$
Gain bandwidth	15 nm

Other Laser-active Dopants in YAG

In addition to Nd:YAG, there are several YAG gain media with other laser-active dopants:

- Ytterbium – Yb:YAG emits typically at either 1030 nm (strongest line) or 1050 nm (→ ytterbium-doped gain media). It is often used in, e.g., powerful and efficient thin-disk lasers.

- Erbium – Pulsed Er:YAG lasers, often lamp-pumped, can emit at 2.94 µm and are used in, e.g., dentistry and for skin resurfacing. Er:YAG can also emit at 1645 nm and 1617 nm.

- Thulium – Tm:YAG lasers emit at wavelengths around 2 µm, with wavelength tunability in a range of ≈ 100 nm width.

- Holmium – Ho:YAG emits at still longer wavelengths around 2.1 µm. Q-switched Ho:YAG lasers are used e.g. to pump mid-infrared OPOs. There are also holmium-doped laser crystals with codopants, e.g. Ho:Cr:Tm:YAG.

- Chromium – Cr^{4+}:YAG lasers emit around 1.35–1.55 µm and are often pumped with Nd:YAG lasers at 1064 nm. Their broad emission bandwidth makes them suitable for generating ultrashort pulses. Note that Cr^{4+}:YAG is also widely used as a saturable absorber material for Q-switched lasers in the 1-µm region.

Neodymium- or ytterbium-doped YAG lasers in the 1-µm region in conjunction with frequency doublers are often the basis of green lasers, particularly when higher powers are required than with directly green-emitting lasers.

Nd: YAG Laser Construction

Nd:YAG laser consists of three important elements: an energy source, active medium, and optical resonator.

Energy Source

The energy source or pump source supplies energy to the active medium to achieve population inversion. In Nd: YAG laser, light energy sources such as flashtube or laser diodes are used as energy source to supply energy to the active medium. In the past, flashtubes are mostly used as pump source because of its low cost. However, nowadays, laser diodes are preferred over flashtubes because of its high efficiency and low cost.

Active Medium

The active medium or laser medium of the Nd:YAG laser is made up of a synthetic crystalline material (Yttrium Aluminum Garnet (YAG)) doped with a chemical element (neodymium (Nd)). The lower energy state electrons of the neodymium ions are excited to the higher energy state to provide lasing action in the active medium.

Optical Resonator

The Nd:YAG crystal is placed between two mirrors. These two mirrors are optically coated or silvered. Each mirror is silvered or coated differently. One mirror is fully silvered whereas, another mirror is partially silvered. The mirror, which is fully silvered, will completely reflect the light and is known as fully reflecting mirror.

On the other hand, the mirror which is partially silvered will reflect most part of the light but allows a small portion of light through it to produce the laser beam. This mirror is known as a partially reflecting mirror.

Working of Nd:YAG Laser

Nd: YAG laser is a four-level laser system, which means that the four energy levels are involved in laser action. The light energy sources such as flashtubes or laser diodes are used to supply energy to the active medium. In Nd:YAG laser, the lower energy state electrons in the neodymium ions are excited to the higher energy state to achieve population inversion.

Consider a Nd:YAG crystal active medium consisting of four energy levels E_1, E_2, E_3, and E_4 with N number of electrons. The number of electrons in the energy states E1, E_2, E_3, and E_4 will be N_1, N_2, N_3, and N_4.

Let us assume that the energy levels will be $E_1 < E_2 < E_3 < E_4$. The energy level E_1 is known as ground state, E_2 is the next higher energy state or excited state, E_3 is the metastable state or excited state and E_4 is the pump state or excited state. Let us assume that initially, the population will be $N_1 > N_2 > N_3 > N_4$.

When flashtube or laser diode supplies light energy to the active medium (Nd:YAG crystal), the lower energy state (E_1) electrons in the neodymium ions gains enough energy and moves to the pump state or higher energy state E_4.

The lifetime of pump state or higher energy state E_4 is very small (230 microseconds (µs)) so the electrons in the energy state E_4 do not stay for long period. After a short period, the electrons will fall into the next lower energy state or metastable state E_3 by releasing non-radiation energy (releasing energy without emitting photons).

The lifetime of metastable state E_3 is high as compared to the lifetime of pump state E_4. Therefore, the electrons reach E_3 much faster than they leave E_3. This results in an increase in the number of electrons in the metastable E_3 and hence population inversion is achieved.

After some period, the electrons in the metastable state E_3 will fall into the next lower energy state E_2 by releasing photons or light. The emission of photons in this manner is called spontaneous emission.

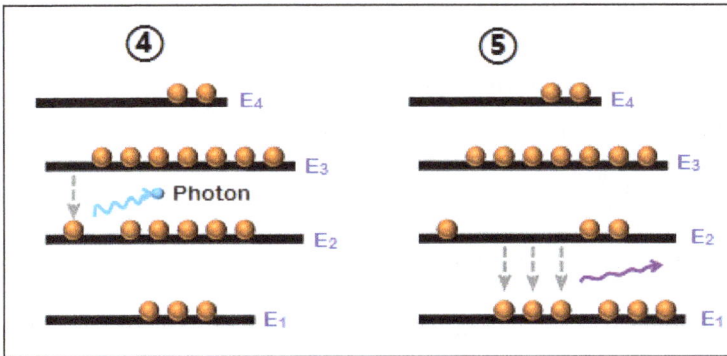

The lifetime of energy state E_2 is very small just like the energy state E_4. Therefore, after a short period, the electrons in the energy state E_2 will fall back to the ground state E_1 by releasing radiationless energy.

When photon emitted due to spontaneous emission is interacted with the other metastable state electron, it stimulates that electron and makes it fall into the lower energy state by releasing the photon. As a result, two photons are released. The emission of photons in this manner is called stimulated emission of radiation.

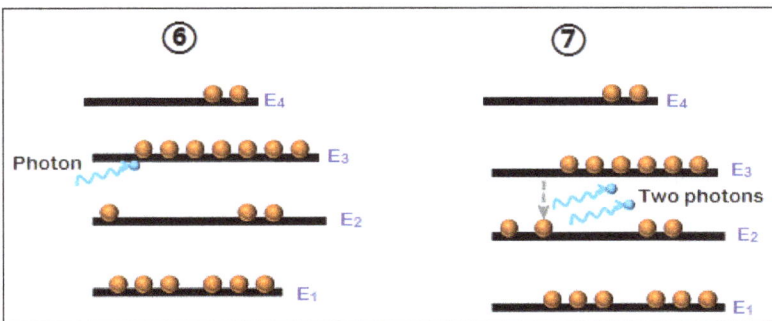

When these two photons again interacted with the metastable state electrons, four photons are released. Likewise, millions of photons are emitted. Thus, optical gain is achieved.

Spontaneous emission is a natural process but stimulated emission is not a natural process. To achieve stimulated emission, we need to supply external photons or light to the active medium.

The Nd:YAG active medium generates photons or light due to spontaneous emission. The light or photons generated in the active medium will bounce back and forth between the two mirrors. This stimulates other electrons to fall into the lower energy state by releasing photons or light. Likewise, millions of electrons are stimulated to emit photons. The light generated within the active medium is reflected many times between the mirrors before it escapes through the partially reflecting mirror.

Advantages of Nd:YAG laser:

- Low power consumption.
- Nd:YAG laser offers high gain.
- Nd:YAG laser has good thermal properties.
- Nd:YAG laser has good mechanical properties.
- The efficiency of Nd:YAG laser is very high as compared to the ruby laser.

Applications of Nd:YAG Laser

Military: Nd:YAG lasers are used in laser designators and laser rangefinders. A laser designator is a laser light source, which is used to target objects for attacking. A laser rangefinder is a rangefinder, which uses a laser light to determine the distance to an object.

Medicine:

- Nd: YAG lasers are used to correct posterior capsular opacification (a condition that may occur after a cataract surgery).
- Nd:YAG lasers are used to remove skin cancers.

Manufacturing:

- Nd:YAG lasers are used for etching or marking a variety of plastics and metals.
- Nd:YAG lasers are used for cutting and welding steel.

Diode Pumped Solid State Laser

A Solid-state laser which made by pumping a solid gain medium is called Diode Pumped Solid State Laser.

Pumping is usually performed in the following forms: (i) optical pumping uses either cw or pulsed light emitted by a powerful lamp or a laser beam. Optical pumping can be realized by light from powerful incoherent sources. The incoherent light is absorbed by the active medium so that the atoms are pumped to the upper laser level. This method is especially suited for solid state or liquid lasers whose absorption bands are wide enough to absorb sufficient energy from the wide band incident incoherent light sources. (ii) electrical pumping is used for gas and semiconductor lasers. It is realized by allowing a current (continuous direct current, radio frequency current or pulsed current) to flow through a conductive medium, such as an ionized gas or semiconductor. Electrical pumping is usually performed by means of sufficiently intense electrical discharge. Gas lasers commonly use electrical pumping or laser pumping, because their absorption bands are narrower than solid and liquid lasers, wide band lamp light is not efficient enough, much of the lamp energy is dissipated as heat. Electrical pumping is non-resonant pumping by electron impact excitation. Electrical pumping is efficient for gases and semiconductors, whose absorption bandwidth is wide enough. Although some optical pumping methods for semiconductor medium have been developed, electrical pumping for semiconductor lasers proved to be more convenient. (iii) chemical pumping, the population inversion is produced directly by exothermic chemical reaction. Chemical pumping usually applies to materials in gas phase, and generally requires highly reactive and often explosive gas mixtures. The exothermic reaction usually generates large amount of energy, if quite a fraction of this available energy is transferred into laser energy, high power and high energy pulses for lasers can be realized. Such lasers are used as directed energy weapons.

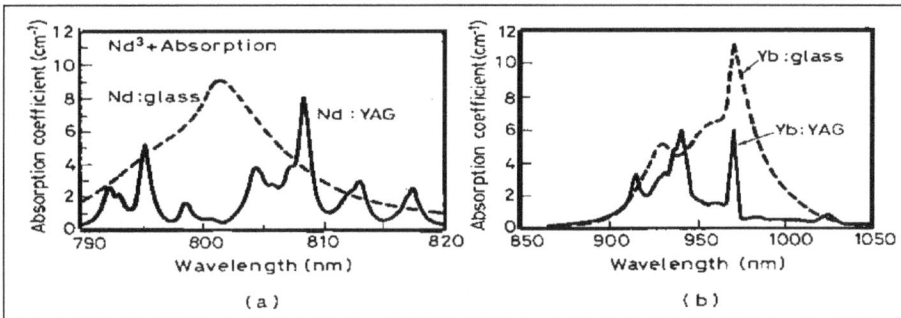

Absorption spectral of (a) Nd:YAG/glass lasers and (b) Yb:YAG/glass lasers.

There are other pumping processes such as gas dynamic pumping, etc. Laser pumping has been used since the early days of the development of lasers. Laser pumping has become a very important pumping technique since efficient and high power diode lasers have been developed and widely available in many wavelengths. When we use diode lasers to pump other solid state lasers, we can produce an all solid state laser. Because optical pumping is a resonant process, the wavelengths of the pumping diode lasers must be within the absorption bandwidth of the active medium to be pumped, the nearer to the absorption peak wavelength the better. Figure shows the absorption spectral of Nd:YAG laser, Nd:glass laser, Yb:YAG laser and Yb:Glass laser. Nd:YAG has a peak absorption value at 810 nm, Nd:glass has a peak value at 802 nm, they can be pumped by GaAs/AlGaAs quantum well diode lasers at about 800 nm. While for Yb:YAG laser and Yb:glass laser, the best absorption wavelengths are 960 and 980 nm respectively, we can pump them using InGaSa/GaAs strained quantum well lasers in the 950-980 nm range. We can divide diode laser pumping into four types according to the degree of integration of the diode lasers: single stripe, diode array, diode bar and diode stack. Normally the pumping power increases with the integration degree.

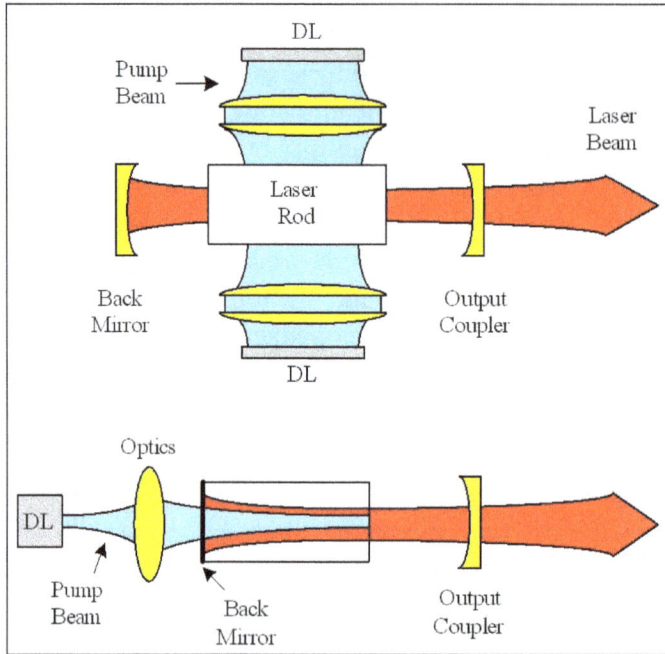

Longitudinal diode laser pumping.

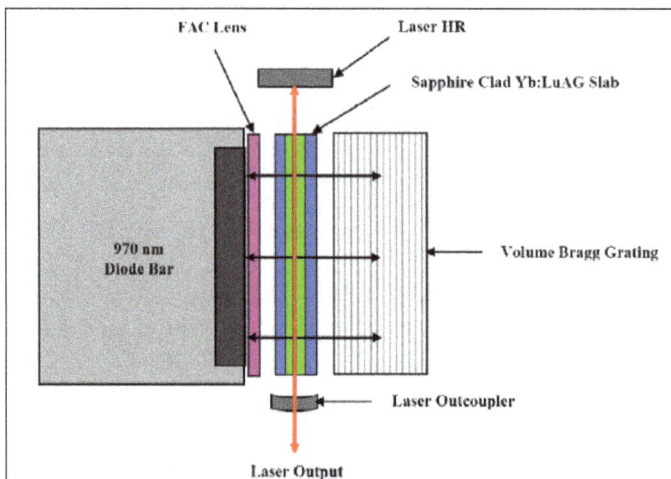

Transverse diode laser pumping.

There are basically two types of pump geometry, longitudinal pumping (pump beam enters the laser medium along the resonator axis) and transverse pumping (pump beam incident on the active medium from transverse directions to the resonator axis). For longitudinal pumping, the beam needs to be concentrated to a small and circular spot. The simplest, if not least costly way to double the pump power is to obtain a higher power pump diode. However, this normally means that the emitting area (stripe width) also increases so that all other factors being equal, the spot or mode size in the laser crystal also increases. One way to double the pump power without increasing the spot size is to optically combine two similar pump diodes. Since these types of edge emitting laser diodes are polarized, a pair of them can be combining using a polarizing beam splitter producing a result that is very nearly double the output power of a single diode, but is non-polarized.

Combining optics consists of:

1. Fast axis correction (optional, cylindrical microlens).

2. Beam collimation (spherical positive lens).

3. Slow axis correction (anamorphic prism pair).

4. Turning mirror or other means for aligning the two beams.

5. Polarizing beam splitter used as beam combiner (PBS cube).

6. Focusing lens(es) if required.

Items 2, 4, and 6 will need to either be adjustable using precision mounts, or be glued in place once positioned properly. The optics and beam splitter must be coated for the desired wavelength. DPSS lasers offer several advantages over the broadband pumping schemes that use cw or pulsed pump sources. The growth in the utilization of diode lasers such as diode arrays or bars to pump solid state lasers results directly from a large volume production of diode lasers and arrays, which have reduced the cost of delivered power from semiconductor laser diode. Also, diode laser characteristics such as wavelength stability, overall efficiency, and operational lifetime (10.000 hours or more) have been significantly improved during the last decade.

Advantages of Diode Pumping

Some of the typical characteristics of DPSS lasers:

Optical efficiency: DPSS lasers are highly efficient because of the direct excitation of the pump beam into the useful absorption band of the lasing ion. Direct excitation minimizes the unwanted losses in the lasing crystal with optical to optical efficiency of up to 70%. Selecting the composition of the host material enables laser diodes to be constructed with wavelengths between 600 nm and approximately 30 μm in the infrared. Over much of this region the output power and lifetime of laser diodes is remarkably poor. Only in wavelength regions where strong commercial or military interest is present have sufficient resources been deployed to develop an efficient and long lived device. Notable examples of this process have occurred at 1.5 μm and 1.3 μm, which are important telecommunications wavelengths, and near 800 nm which is important for communications, entertainment and medical applications. Only in the vicinity of 800 nm have high powered devices been built. The driving force has been military applications including pumping Nd doped solid state lasers.

Wavelength: The wavelength at which laser diodes operate is dictated by the size of the band gap, since the light arises from the recombination of electrons and holes in a pn junction. The band gap may be tuned in size by two main processes: (i) Altering the composition of the host material. (ii) Changing the temperature of the host material. Other physical effects such as the application of pressure may also change the band gap but they tend to produce too small an effect to be useful. The output wavelength of diode lasers varies from diode to diode because of small differences in fabrication and the wavelength changes with temperature. The variation in output wavelength leads to increased cost because only diode lasers in a small wavelength range are usable. The change in wavelength resulting from temperature variation requires that the diodes must be temperature controlled.

Operational lifetime: The operational lifetime of laser diodes or arrays is much larger than that of conventional arc or filament lamps. A typical laser diode array can operate without significant degradation for more than 10.000 hours, but usually up to 3 x 104 hours, while a cw lamp must be replaced after 200-400 hours of operation (or 107 shots in the case of pulsed pumping). The performance of a diode laser degrades exponentially with time. Initially, the failure rate is low, but it increases exponentially with the operating time. Failure mechanisms of laser diodes are divided into two main classes:

- User induced damage (such as mechanical, thermal or electrical shock or electrostatic discharge).

- Intrinsic damage, which results from three main sources: (i) degradation of laser mirrors or facets because of high current densities or current spikes. This will increase internal losses and lead to catastrophic failure. (ii) damage resulting from crystal defects within the active region, which will lead to an increase in absorption losses within laser diode. Such defects are common in AlGaAs laser diodes because of the oxidation and migration of the aluminum. At present efficient aluminum free laser diodes replace the aluminum containing lasers. (iii) resistive losses and heating because of increases in the resistance of electrical contacts to the laser.

Temperature: Since the diode laser is narrow bandwidth pumping source, it pumps only the useful absorption bands relevant to laser action, reducing the thermal load in the crystal. This thermal load results from the quantum gap between the pump and the leasing photons. Thermal effects such as thermal lensing, thermally induced birefringence, and thermal damage to the lasing to the crystal are reduced significantly. Fine tuning the temperature of the laser diode causes a change in wavelength. For GaAlAs devices the wavelength tunes at an approximate rate of +0.25 nm $°C^{-1}$ mainly due to the change in band gap with temperature. This feature is used to tune the laser diode into coincidence with the absorption bands of rare earth ions. Normally the lasers will be cooled down, since this gives a longer lifetime for the laser diode. It is usual to specify the room temperature wavelength of the laser diode some 5 nm longer than the rare earth absorption feature that will be pumped. Simple Peltier coolers can provide a temperature change of about 40 °C corresponding to a wavelength shift of 10 nm. The Nd:YAG absorption linewidth is ~2 nm so the diode temperature has to be set to better than ±4 °C, which is not too demanding in the laboratory. The influence of temperature on the wavelength of the laser diode causes problems of packaging the DPSS laser. Pulsed laser diodes suffer from a transient thermal wavelength shift. Since the current is pulsed through the laser diode, the temperature is never in equilibrium and a transient wavelength shift occurs. The wavelength increases during the optical pulse. Measurements of this effect have revealed shifts of ~5 nm. Shifts of this magnitude are larger than the absorption linewidth of many solid state laser materials. Account has to be taken of this effect when predicting the efficiency of pulsed DPSS lasers. The diode lasers are normally cooled by thermoelectric cooler for low power systems, and by liquid cooling for high powers. Cooling can stabilize the diode laser frequency.

Beam quality: Although the beam quality of a laser diode or diode array is not good, the use of coupling optics makes it possible to obtain a good TEM00 beam mode form a DPSS laser. The coupling optics circularize the output beam emanating from the laser diode array or bar, and then couple the beam into the solid state laser crystal either by direct coupling or an optical fiber. The cylindrical fast axis collimating lens can reduce the beam divergence of diode laser stacks to a value of lens than 10 mrad. The solid state laser can be pumped longitudinally or transversely. The laser crystal host can be in the form of a crystal, waveguide or optical fiber.

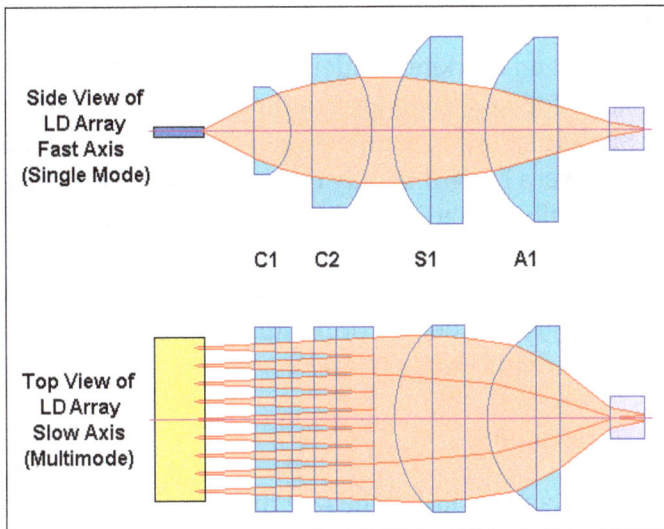

The elliptic beam shape arising from the transverse dimensions of a single laser diode stripe.

Since the absorption length of the diode laser beam focused inside the solid state laser crystal is short, the pump mode volume is smaller than the laser cavity mode volume and one expects a good spatial beam quality. The laser cavity itself is short and therefore the output power of the DPSS laser is a single longitudinal mode. The best transverse mode quality is obtained from single stripe devices. The low ellipticity of a source with dimensions of 3 x 1 μm ensures that the output power may be efficiently coupled into optical fibre or into solid state laser materials. Single laser diodes may provide up to 150 mW in a single transverse mode. The shape of the output beam from a single strip device is shown in figure.

To date the most extensively used pumping geometry is longitudinal pumping. The output from the diode laser is collimated and beam shaped to achieve a circular profile before being focused down to form a pump spot on the laser rod. This technique allows good matching between the lasing spot size of the solid state laser cavity and the pump spot size in the gain medium. This spatial overlap between pump and lasing modes, known as mode matching, is critical to the efficiency of the diode pumping process.

Frequently, DPSS lasers are pumped by laser diode arrays in order to achieve high output powers. As has been noted, the simplest way to increase the output power from a single laser diode is to increase the width of the emitting region either by fabricating a number of laser diode stripes in close proximity so that there is a series of emitting regions, or by enlarging the width of the electrically pumped region. The output of arrays lies in the region of 200 mW to 3 W cw output at present. True arrays of laser diode stripes tend to be partially coherent and may exhibit the two lobe structure in the far field characteristic of the phase changes due to evanescent coupling between adjacent stripes. This feature appears less obvious as the output power and number of stripes increase due to reduced coherence across the array. The broad stripe arrays are multi transverse mode devices and the beam quality depends on how hard they are driven. Applications of high power fiber delivery from diode lasers include pumping either bulk or fiber lasers, machining and marking, soldering, and power transmitting. High power fiber coupled diode laser sources could lead to the creation of compact high power diode pumped lasers. Similar systems that have been demonstrated either have relatively low power because they were limited to only one or two sources focused into a fiber, or because they consist of a fiber bundle in which one or two sources were focused into a fiber and then many fibers were brought together to form the pump source. A single fiber to carry all the power, as opposed to a fiber bundle, is desirable because it is less bulky and it maintains a greater degree of pump beam brightness, which allows improved performance from diode pumped lasers.

Common DPSS Processes

The most common DPSS laser in use is the 532 nm wavelength green laser pointer. A powerful (>200 mW) 808 nm wavelength infrared GaAlAs laser diode pumps a neodymium doped Nd:YAG or a neodymium doped yttrium orthovanadate (Nd:YVO$_4$) crystal which produces 1064 nm wavelength light from the main spectral transition of neodymium ion. This light is then frequency doubled using a nonlinear optical process in a potassium titanyl phosphate (KTiOPO$_4$, KTP) crystal, producing 532 nm light. Green DPSS lasers are usually around 20% efficient, although some lasers can reach up to 35% efficiency. In other words, a green DPSS laser using a 2.5 W pump diode would be expected to output around 500-900 mW of 532 nm light. In optimal conditions,

Nd:YVO$_4$ has a conversion efficiency of 60%, while KTP has a conversion efficiency of 80%. In other words, a green DPSS laser can theoretically have an overall efficiency of 48%. In the realm of very high output powers, the KTP crystal becomes susceptible to optical damage. Thus, high power DPSS lasers generally have a larger beam diameter, as the 1064 nm laser is expended before it reaches the KTP crystal, reducing the irradiance from the infrared light. In order to maintain a lower beam diameter, a crystal with a higher damage threshold, such as lithium triborate (LBO), is used instead. Much of the excitement in nonlinear optics has been caused by a new generation of nonlinear materials. These new materials are non-hygroscopic, have high damage threshold and good phase matching characteristics. Materials which are now widely available include KTP, LBO potassium niobate (KNB) and beta barium borate (BBO).

Blue DPSS lasers use a nearly identical process, except that the 808 nm light is being converted by an Nd:YAG crystal to 946 nm light (selecting this non principal spectral line of neodymium in the same Nd doped crystals), which is then frequency doubled to 473 nm by a BBO or LBO crystal. Because of the lower gain for the materials, blue lasers are relatively weak, and are only around 3-5% efficient. In the late 2000s, it was discovered that bismuth triborate (BiBO) crystals were more efficient than BBO and LBO and do not have the disadvantage of being hygroscopic, which degrades the crystal if it is exposed to moisture.

Violet DPSS lasers at 404 nm have been produced which directly double the output of a 1.000 mW 808 nm GaAlAs pump diode, for a violet light output of 120 mW (12% efficiency). These lasers outperform 50 mW gallium nitride (GaN) direct 405 nm Blu ray diode lasers, but the frequency doubled violet lasers also have a considerable infrared component in the beam, resulting from the pump diode. Yellow DPSS lasers use an even more complicated process: A 808 nm pump diode is used to generate 1,064 nm and 1,342 nm light, which is summed to become 593.5 nm. Due to their complexity, most yellow DPSS lasers are only around 1% efficient, and usually more expensive per unit of power. In addition to optical fibre waveguides, waveguide lasers have been fabricated in bulk glasses and in a variety of crystals.

Examples of DPSS Lasers: We consider the performance of a range of specific DPSS lasers. We begin by considering in some detail the laser diode pumped Nd:YAG laser, which is certainly the best studied laser diode pumped system. We will then consider the performance of the laser diode pumped Nd:X laser, where X is any one of a number of host materials and also laser diode pumped stoichimetric materials. The stimulus for the research into these materials is essentially to reduce the size of DPSS lasers.

Nd Doped DPSS Lasers

The trivalent neodymium ion was the first of the rare earth ions to be used in a laser and it is the dopant which has received most attention in the field of DPSS lasers. The Nd^{3+} ion has a strong absorption at 0.81 μm, which coincides with the emission wavelength of commercially available GaAs and GaAlAs laser diodes. Most of the research reported on laser diode pumped Nd doped lasers has concentrated on the four level high gain $^4F_{3/2}$ - $^4I_{11/2}$ transition which corresponds to a laser output in the region of 1.06 μm. However, work has also been reported on the 1.3 μm $^4F_{3/2}$ - $^4I^{13/2}$ transition, and on the three level 0.946 μm $^4F_{3/2}$ - $^4I_{9/2}$ transition. All Nd doped laser results refer to the 1.06 μm transition unless otherwise stated. Figure shows a simplified energy level diagram for Nd and several other important rare earth ions.

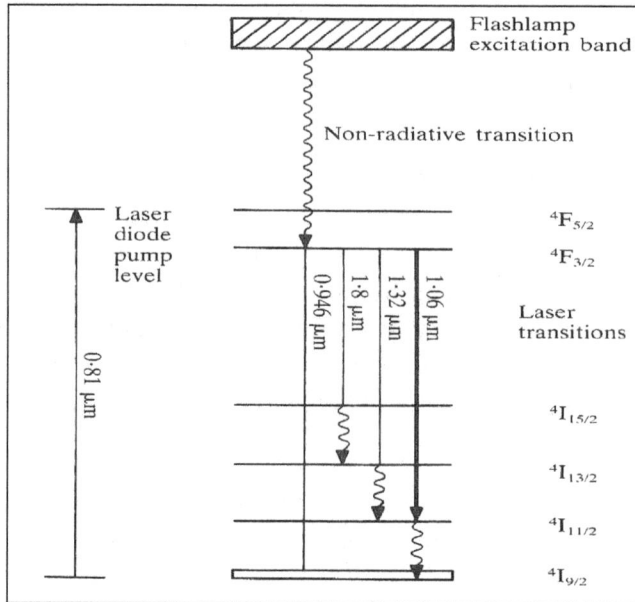

Enegy level diagram for Nd:YAG, illustrating process
of diode laser excitation and Nd:YAG laser transitions.

The upper state lifetime is another important parameter for energy storage which can vary significantly with the choice of host medium. The problem which effects Nd^{3+} doped laser materials is that of concentration quenching of the upper state lifetime, where the lifetime decreases with increasing Nd^{3+} concentration. This leads to an increase in pump threshold. However, high doping is often desirable in DPSS lasers so that the pump radiation can be absorbed in a smaller volume, yielding a lower pump power threshold. One class of material in which this problem may be overcome is that of stoichiometric materials, where the Nd^{3+} is a constituent component of the material, rather than a dopant.

Laser Diode Pumped Nd:YAG Laser

The Nd:YAG laser has become the most common DPSS laser for a variety of reasons. The advantages of the Nd^{3+} ion as a dopant have been mentioned previously. The level of neodymium doping in YAG is limited to about 1.5% due to concentration quenching of the upper state lifetime. The YAG host is hardness, high thermal conductivity and good optical quality.

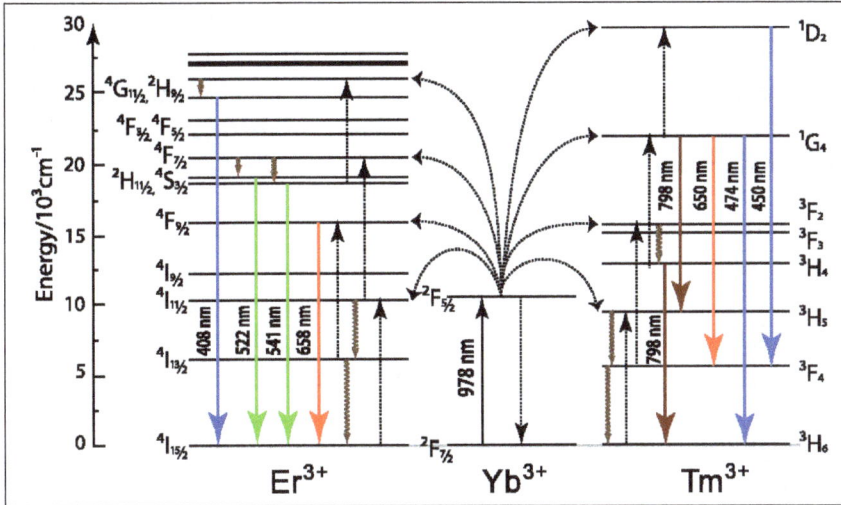

Simplified energy level diagrams for some
of the most important rare earth ions.

The highest overall efficiencies are generally obtained from the longitudinally or end pumped geometry, due to the excellent matching that can be obtained between the solid state laser TEM$_{00}$ mode. The first highly efficient laser diode end pumped Nd:YAG laser was reported by Sipes, who obtained an output of 80 mW at an overall (electrical to optical) efficiency of 8%. The limitations imposed on efficient coupling using end pumping mean that to obtain higher powers it is necessary to use a scheme whereby a laser rod or slab is pumped using a transverse pumping geometry. A disadvantage of the transverse pumped geometry tend to be much lower than for the end pumped case due to poor mode matching. Optical slope efficiencies of 23% for slabs and 54% for rods have been obtained.

Transversely pumped cw Nd:YAG lasers have also been reported. Burnham and Hays used four cw laser diodes to pump a Nd:YAG rod transversely. 3.3 W of multimode output were obtained, at an overall efficiency of 3.5%. As expected, the efficiency of TEM00 operation was lower at 2%.

Laser Diode Pumped Nd:YLF Laser

The Nd:YLF laser has several properties very different from those of the Nd:YAG laser. The fact that it is a uniaxial material means that, simply by using an intracavity polarizer, one can select one of two different wavelengths for each transition. For the $^4F_{3/2} - \,^4I_{11/2}$ transition these wavelengths are 1.047 µm, and 1.053 µm, with the 1.047 µm polarization exhibiting the higher gain. The natural birefringence of Nd:YLF swamps the effect of thermally induced birefringence observed in materials such as Nd:YAG.

The fluorescence lifetime of Nd:YLF is approximately twice as long as that of Nd:YAG. Furthermore, a large amount of energy can be stored in the medium gain 1.053 µm transitions before the onset of amplified spontaneous emission. These facts make the laser diode pumped Nd:YLF laser an attractive medium for the generation of high power Q switched pulses. The results show that approximately twice the pulse energy can be obtained from a Q switched Nd:YLF laser than from a Q switched Nd:YAG laser, due to the difference in the fluorescence lifetime for these media. The highest peak power that has been obtained to date from the laser diode pumped Nd: YLF laser is 70 kW, in a pulse of duration.

Q switching was demonstrated in light emitting diodes pumped Nd:YAG and Nd doped potassium gadolinium tungstenate (Nd:PDT). A number of different techniques for Q switching were used including electrooptic, acoustooptic, saturable absorber, and cavity dumping. Pulse lengths as short as 4 ns using cavity dumping were obtained, with the highest peak power being 170 W in a 65 ns pulse in Nd:PGT by acoustooptic mode locking.

Laser Diode Pumped Nd: Glass Laser

Nd:glass is an ideal candidate for laser diode pumping due to its wide absorption spectrum in the region of 800 nm. Unlike, for example, the YAG host, the concentration of active ions can be very high before the onset of concentration quenching (approximately 7% Nd_2O_3 in phosphate glasses). The main disadvantage of glass as a host medium is its low thermal conductivity (0.6 W $m^{-1}K^{-1}$ for Schott LG760 phosphate glass as opposed to 11 W $m^{-1}K^{-1}$ for Nd:YAG). This makes the Nd:glass laser particularly susceptible to thermal effects, such as thermal lensing, thermally induced birefringence and thermal damage. The first reported operation of an laser diode pumped Nd:glass laser was by Kozlovsky et al. A pump power threshold of 2 mW and a slope efficiency of 42% was obtained from a monolithic device, when pumped with a low power single stripe laser diode. The highest output power obtained from an laser diode pumped Nd:glass laser was reported by Fan. Using a spinning glass disc as the gain medium to overcome thermal problems, an output power of 550 mW was obtained for an absorbed power of 2 W. Basu and Byer carried out an analysis of the scalability of laser diode pumped Nd:glass lasers in the zig zag slab and rotating disc geometries. They predicted that for the rotating disc laser, up to 20 kW of average output power should be achievable.

Diode Pumped Stoichiometric Lasers

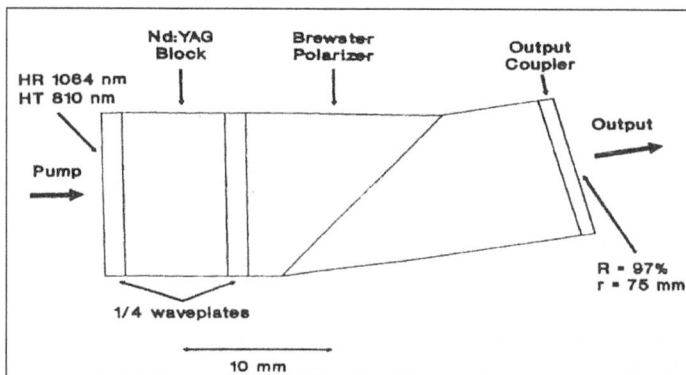

Schematic diagram of the monolithic twisted mode cavity Nd:YAG laser.

We consider the work carried out to date on stoichiometric laser materials. The stimulus for the research into these materials is essentially to reduce the size of DPSS lasers. Stoichiometric materials are compounds which contain the lasing species, rather than hosts into which the species is doped. Due to this, the concentration of active ions can be much larger in stoichiometric materials, resulting in a shorter absorption length for the pump radiation.

Dixon et al. have used an interesting technique to couple the pump radiation into the LNP ($LiNdP_4O_{12}$) laser mode where the crystal is placed in very close proximity to the laser diode output facet. Due the very short absorption length of the LNP crystal, there is no need for collimating and focusing optics to be used, as is normally the case for DPSS lasers. Using a separate output coupler, laser operation was obtained at both 1047 nm (slope efficiency 33%) and 1317 nm (slope efficiency 10%). The maximum output powers obtained were 73.5 mW and 24 mW, respectively. Using a monolithic device, 28 mW output was obtained at 1317 nm. The slope efficiencies reported will increase as crystal growth and fabrication techniques are improved. If the axes of the wave plates are properly adjusted, a standing wave with an axially uniform energy density can be created. Spatial hole burning is thus eliminated.

Diode Pumped Waveguide Lasers

In addition to bulk lasers, there is much interest in the use of laser diodes to pump waveguide lasers. The use of guided wave structures instead of bulk devices is also of interest for diode pumping. One reason is the compatibility of guided wave devices with optical fiber systems. Another reason is that by guiding both the pump wave and the laser mode, higher pump densities and therefore gain compared to bulk devices can be achieved. The first report of a guided wave device was end pumped a multimode, Nd doped, silica based, glass fiber laser. The cavity was provided by either depositing reflector coating directly onto the end of the fiber or by external mirrors.

Waveguiding structures were also considered for a transverse pump geometry. Such a device may offer advantages such as small resonator construction and good transverse mode stability. Design calculations were performed for an light emitting diode array pumped, rectangular cross section waveguide fabricated of a number of Nd laser materials including Nd:YAG, and LNP. Such a device made of an LNP gain medium with a glass cladding was demonstrated using Ar^+ laser pumping.

Fiber Laser

A fiber laser is a laser where the active medium being used is an optical fiber that has been doped in rare elements; typically, erbium, ytterbium, neodymium, thulium, praseodymium, holmium or dysprosium. While you don't need to worry too much about which rare-earth elements have been used, the main thing to note is that it is fiber that is being used at the centre of this laser machine.

This is different to the two other main types of laser, which are gas lasers (typically uses helium-neon or carbon dioxide) and crystal lasers (using Nd:YAG). Fiber lasers are the newest type of laser to hit the market, with many arguing that it is the more beneficial of the three types.

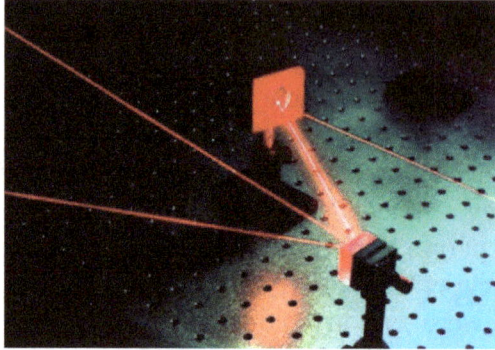
An example of a crystal laser.

How does a Fiber Laser Work?

The fiber used as the central medium for your laser will have been doped in rare-earth elements, and you will most often find that this is Erbium. The reason this is done is because the atom levels of these earth elements have extremely useful energy levels, which allows for a cheaper diode laser pump source to be used, but that will still provide a high output of energy.

For example, by doping fiber in Erbium, an energy level that can absorb photons with a wavelength of 980nm is decayed to a meta-stable equivalent of 1550nm. What this means is that you can use a laser pump source at 980nm, but still achieve a high quality, high energy and high power laser beam of 1550nm.

The internal Bragg Grating acts like a set of mirrors inside the core.

The Erbium atoms act as the laser medium in the doped fiber, and the photons that are emitted remain within the fiber core. To create the cavity in which the photons remain entrapped, something known as a Fiber Bragg Grating is added.

A Bragg Grating is simply a section of glass which has stripes in it – which is where the refractive index has been altered. Any time that light passes across a boundary between one refractive index and the next, a small bit of light is refracted back. Essentially, the Bragg Grating makes the fiber laser act like a mirror.

The pump laser is focused into cladding that sits around the fiber core, as the fiber core itself is too small to have a low-quality diode laser focused into it. By pumping the laser into the cladding around the core, the laser is bounced around inside, and every time that it passes the core, more and more of the pump light is absorbed by the core.

Usefulness of Fiber Laser

While the above may have been a bit science-heavy, we thought we'd share with you some of the benefits that come with the way that a fiber laser works. One of the biggest benefits that a fiber laser offers to its users is that it is extremely stable.

Other normal lasers are very sensitive to movement, and should they get knocked or banged, the whole laser alignment will be thrown off. If the optics themselves get mis-aligned, then it can require a specialist to get it working again. A fiber laser, on the other hand, generates its laser beam on the inside of the fiber, meaning that sensitive optics aren't required to have it working properly.

Another huge benefit in the way that a fiber laser works is that the beam quality that is delivered is extremely high. Because the beam, as we've explained, remains contained within the core of the fiber, it keeps a straight beam that can be ultra-focused. The dot of the fiber laser beam can be made incredibly small, perfect for applications such as laser cutting.

While the quality remains high, so too does the level of power that the fiber laser beam delivers. The power of a fiber laser is constantly being improved and developed, and we now stock fiber lasers that have a power output over 6kW (number 15). This is an incredibly high level of power output, especially when it is super focused, meaning it can easily cut through metals of all kinds of thicknesses.

Another useful aspect in the way in which fiber lasers work is that despite their high intensity and high power output, they are extremely easy to cool while remaining highly efficient at the same time. Many other lasers will typically only convert a small amount of the power that it receives into a laser. A fiber laser, on the other hand, converts somewhere between 70%-80% of the power, which has two benefits.

The fiber laser will remain efficient by using near-to 100% the input that it receives, but it also means that less of this power is being converted into heat energy. Any heat energy that is present is evenly distributed along the length of the fiber, which is usually quite long. By having this even distribution, no part of the fiber gets too hot to the point where it causes damage or breaks.

Finally, you'll also find that a fiber laser works with low amplitude noise, is also extremely resistant to heavy environments, and has low maintenance costs. Generally, an SPI fiber laser will require no servicing as they are built with our 'fit and forget' technology. However, for the rare occurrence maintenance should ever be required costs are typically around 50% less than other lasers.

Dye Laser

The Dye Laser is a Liquid Laser. Liquid lasers are those lasers which uses liquid as an active medium In dye laser the liquid material called dye (for example rhodamine B, sodium fluoresein and rhodamie 6G) uses as an active medium, which causes to produce laser light.

Characteristics of Dye Lasers

The dye lasers produce output whose wavelengths are in the visible, ultra violet and near infrared spectrum. Which usually depending on the dye used wave lengths therefore vary from 390 to 1000nm.

The output power of dye lasers can be considered to start from 1 watt with no theoretical upward limit. The output beam diameter is typically 0.5mm and the beam divergence is from 0.8 to 2 milli radians. The conversion efficiency of the light from the pumping source to an output from the dye laser si relatively high approximately 25%.

Construction of Dye Liquid Lasers

The dye lasers can be constructed in two possible ways, according to their pumping methods. The 1st configuration can be shown as:

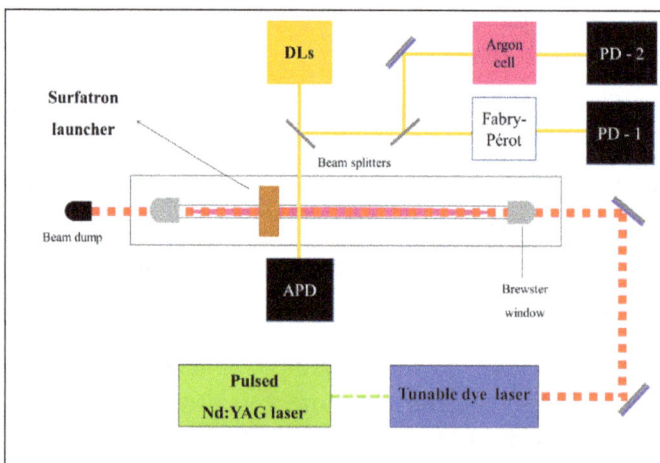

Dye Laser Construction.

In the shown configuration, the dye is pumped through the capillary tube from a storage tank. While in capillary tubes it is optically excited by flash lamp. The output of the laser passes through a Brewster window to the output coupler which is 50% reflective mirror.

The 2nd configuration can be shown as:

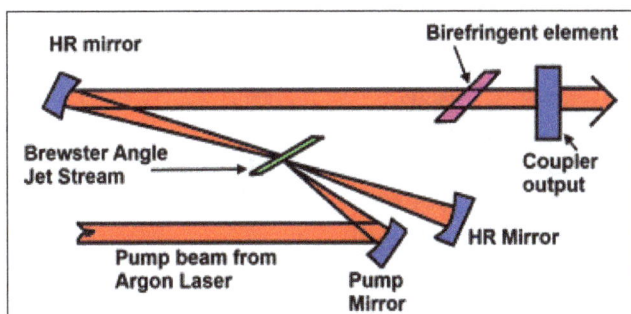

Dye Laser Construction 2nd Configuration.

In this shown configuration, the dye is pumped through the nozzle at high speed to form a Brewster angle jet stream. The excitation mechanism for this laser is a second laser (e.g. argon laser). The laser beam reflected from two HR (high Reflective) mirrors to the output coupler again the output coupler is about 50% reflective. The birefringent filter/element is used to tune the laser to one of a given range of frequencies.

Working/Function of Dye Lasers

We know that active medium used in a dye laser can be one of organic dyes. The medium is dissolved in a solvent such as water, alcohol or ethylene glycol. The organic dyes such as rhodamine B, sodium fluorsin for example chemical formula for one of these dyes rhodamine-B is c28H31. It is therefore very difficult to determine the element that actually lases. For this reason we will simply say that some organic dye will lase.

The organic dye laser produces a range of wavelengths. For example rhodamine-B produces wavelengths in the 590nm to 660nm range. However the amount of amplification varies across the range of frequencies, with max output at about 618nm.

By using the birefringent filter, it is possible to tube the laser to specific output frequency. This filter bends the different wavelengths much the same as a prism but to as much greater extent. This makes it possible to tube the laser with great deal of accuracy ube the laser with great deal of accuracy.

Energy Transfer in Organic Dyes

Organic dye molecules are complicated structures composed of a large number of atoms of several species. For example, the chemical formula for the dye known as rhodamine B is: $C_{28}H_{31}ClN_2O_3$. This indicates that a molecule of this dye contains carbon (C), hydrogen

(H), chlorine (Cl), nitrogen (N), and oxygen (O). The subscript beneath a given symbol denotes the number of atoms that species, e.g., H_{31} represents 31 hydrogen atoms.

The energy-level structure of an organic dye molecule is correspondingly complex. A highly simplified energy-level diagram for a typical dye is shown in figure.

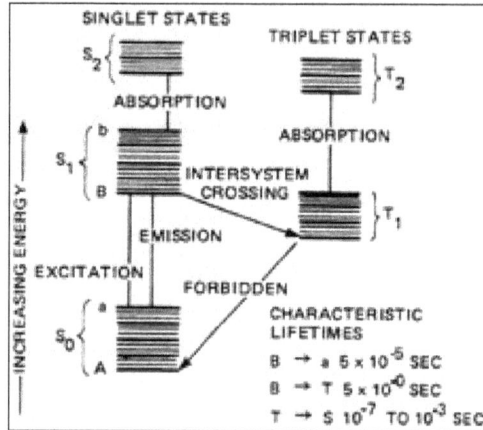

Energy-level diagram of a laser dye.

Two sets of electronic states are displayed—singlets and triplets. Typical energy separation between two singlet or two triplet states is about 10,000-20,000 cm^{-1}. These electronic states are split by vibrational motion of the molecules. The various vibrational sublevels are indicated by heavy horizontal lines in figure. The separation between two neighboring vibrational states is of the order of 1000-2000 cm^{-1}. Rotational molecular motion produces further level splitting (fine structure), indicated in the diagram by the lighter horizontal lines. Two adjacent rotational states are separated by about 10-20 cm^{-1}. The complexity of the dye molecule and the factors leading to line splitting produce broad absorption and emission curves as illustrated in figure.

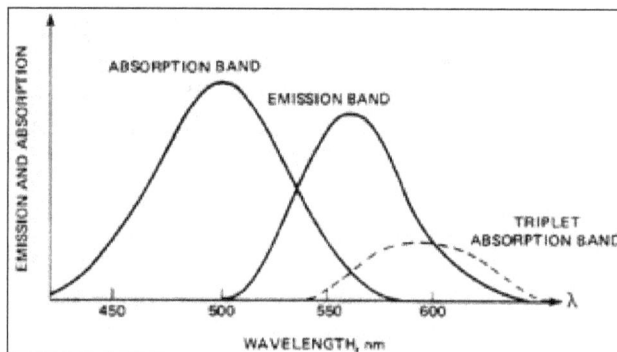

Absorption and emission (fluorescence) spectrum of a typical laser dye.

Lasing Transitions in Dyes

This topic traces out the sequence of steps that lead to lasing in a dye laser. Losses due to population of the T_1 triplet state and T_1 (r) T_2 absorption are neglected initially.

A dye laser can be considered to be basically a four-level system. Pump light absorbed by dye molecules in the lowest vibrational sublevel A of the S_o singlet ground state produces transitions to one of the upper vibrational levels of S_1, as denoted by b in figure. The molecule then undergoes a radiationless decay to the bottom of S_1 (b (p) B). The radiationless transition involves a rearrangement of total energy within a system without photon emission. In most instances, the energy is absorbed in increased kinetic energy or motion of the atoms in the system and appears as heat. Lasing may occur between vibrational sublevel B in S_1 and an excited sublevel such as "a" within S_o, provided a population inversion exists between these two states $N_B > N_a$). The B (p) a transition is accompanied by the emission of a photon of wavelength l VA, as shown in equation below:

$$\lambda_{Ba} = \frac{hc}{E_B - E_a}$$

h = Planck's constant 6.625 × 10⁻³⁴ joule-second).

where: c = Speed of light (3 × 10⁸ meters/second)

Another radiationless transition, a (p) A, returns the excited molecule to its ground state.

The broad absorption and fluorescence bands of a typical dye in solution. The wavelength of light emitted by the dye when it fluoresces is longer than that of the absorped pump light:

$$| E_A - E_b | > | E_B - E_a |$$

or

$$\frac{hc}{E_A - E_b} < \frac{hc}{E_B - E_a}$$

so that

$$\lambda_{Ab} < \lambda_{Ba}$$

or finally

$$\lambda_{Ba} < \lambda_{Ab}$$

where the subscripts indicate the B (p) a emission and A (p) b absorption transitions, respectively. Thus, one observes that absorption of light at a given wavelength is re-emitted (fluorescence) at a longer wavelength. This property of organic dyes is used in a number of familiar commercial products such as brightly colored plastic signs and drafting equipment (triangles, protractors, etc). In detergents, cloth, and paper, certain dyes acts as brightening (whitening) agents by absorbing UV and emitting in the blue portion of the spectrum. The ability to absorb light of short wavelengths and fluoresce or re-emit light of longer wavelengths is one of the most useful properties of organic dyes.

Triplet Absorption in Dyes

The small-signal gain per unit length of a typical dye is ~10^3 cm^{-1}, larger by several orders of magnitude than that of other lasers such as HeNe (10^{-3} cm^{-1}) or ruby (7×10^{-2} cm^{-1}). Thus, it would seem an easy task to make a dye laser system work. However, the "troublesome triplets" are a source of difficulties that diminish the advantages of large gain.

Figure may be used to illustrate the effects of triplet absorption in dyes. Molecules in the upper lasing level B in S_1 may decay to the triplet state T_1 by a radiationless transition (S_1 (p) T_1). This is a forbidden transition in the sense that it is far less probable than a singlet-singlet (S_1 (p) S_0) transition. Nevertheless, molecules can undergo an "intersystem crossing" (S_1 (p) T_1) in a time ~50 nsec. T_1 is a metastable state with a lifetime of about 10^{-7}–10^{-3} sec for a decay (t_1 (p) S_0). Hence, molecules can pile up in T_1, "stealing" power that could go into lasing via B (p) a transitions. A second problem, that of triplet-triplet (T_1 (p) T_2) absorption, is even more serious. Figure illustrates the wavelengths at which this absorption (T_1 (p) T_2) occurs overlap the wavelengths for which lasing is expected. Triplet-triplet absorption can be so strong that the excitation mechanism (CW laser, pulsed laser, or flashlamp) may not have sufficient power to overcome this absorption loss. This can result in no lasing at all or in early termination of lasing. If the population of the triplet state T_1 is not limited by some means, it will set the maximum duration of dye laser pulses at about 0.1 m sec. Thus, uncontrolled triplet-triplet absorption would make long-pulse (> 0.1 m sec) and CW operation of a dye laser impossible.

Triplet Quenching

Triplet absorption in a flashlamp-excited dye systems is believed to be a major factor in the experimentally observed lack of synchronization between the pump pulse and the laser pulse. That is, the laser pulse terminates before the pump pulse ends. In fact, the laser pulse usually terminates before the intensity of the pump pulse has fallen below the threshold excitation value.

Attempts to construct a CW dye laser led to efforts aimed at controlling the triplet state population. One method often used is to add a second molecule to the dye solution to act as a so-called triplet *quenching agent*, which promotes T_1 (p) S_0 transitions. Collisions between quencher and dye molecules are responsible for this de-excitation process. A number of substances have been discovered that have the correct energy-level structure to be effective as triplet quenchers. The one selected depends upon the specific organic dye to be used. For example, a chemical known as cyclooctatetraene (COT) is a good triplet quencher for rhodamine 6G.

In CW dye lasers rapid flow of the dye solution is also used to control triplet state population. The dye flows very rapidly through the active gain region where the argon beam is focused. In flowing through this region each dye molecule "sees" a pulse of pumplight. If the dye is circulated fast enough, the individual dye molecule will be irradiated by a light pulse of duration *short* compared to the triplet decay (T_2 (p) S_0) time but *long* with respect to the lifetime of the lasing transition (B (p) a). As an example, suppose the focused spot diameter of the pump beam is 10 microns. If the dye solution is circulated at a speed $v = 10$ m/sec, then each dye molecule "sees" a light pulse of duration t, where:

$$t = \frac{d \text{ focused beam}}{v(\text{dye speed})}$$

$$= \frac{10^{-5} \text{ m}}{10 \text{ m/sec}}$$

$$t = 1\mu \sec.$$

CW Pumped Dye Lasers

This examines some design problems posed by a CW pumped dye laser and their solutions. The power density necessary to reach threshold for lasing in a typical organic dye laser is fairly large, ~100 kW/cm². This value of irradiance generally precludes the use of CW arc lamps or other incoherent excitation sources. The focused beam from commercially available argon lasers, on the other hand, easily can exceed the required threshold power density. In addition, argon lasers have multiline outputs in a range of wavelengths from the near UV to the blue-green, which is convenient for exciting a reasonably large number of dyes. Thus, argon ion lasers are used almost universally for pumping CW dye lasers.

Basic Design Features

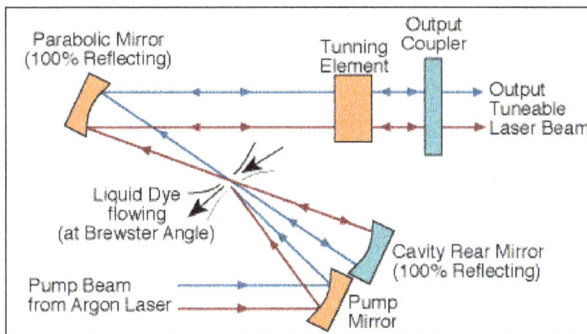

CW dye laser optically pumped with an argon laser.

Figure shows the basic design of an argon-pumped, CW dye laser. The optical cavity of the dye laser consists of three mirrors. The output coupler is a long-radius or flat mirror. Its transmission is typically between 10 and 20 percent. The two high-reflectance mirrors are curved and mounted at the proper separation and alignment to produce a focal point in the beam within the dye jet. Pump light in the form of the argon laser beam is focused into the dye jet at the same point as the dye laser beam.

Figure shows the jet assembly of a CW dye laser. The dye is forced out of a small stainless steel nozzle in a broad, flat stream. The sides of this dye stream are relatively flat over the small area of the focused laser beams and serve as optical surfaces. The jet is oriented at Brewster's angle to minimize surface reflections. To exceed the threshold power density by a factor of 5, say, the beam from a 1-watt argon laser must be focused down to a very small diameter, ~10-20 µm. However, the focused beam can create a "hot spot" in the dye. Thermal gradients in the dye solution then result in optical inhomogeneities—that is, localized changes in the refractive index of the solution—and subsequent distortion of the output beam.

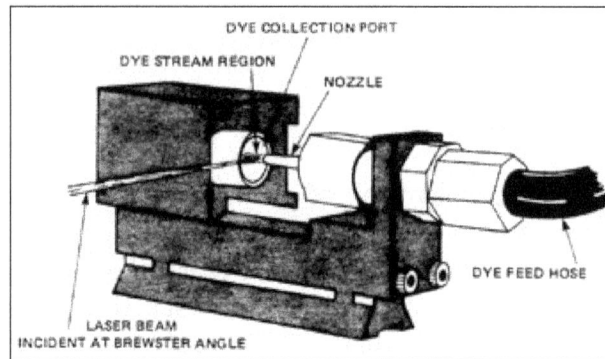

Nozzle assembly of a laminar-flow dye laser.

Inhomogeneities that arise from thermal distortion can be controlled by rapid flow of the dye solution (~10-20 m/sec) and by use of a dye solvent of which the refractive index is relatively independent of temperature, namely, water. This, in addition to convenience, is the reason that water often is used as the solvent for dyes in CW pumped dye lasers.

When water is used as a dye solvent, another problem arises, because most organic dyes do not dissolve readily in water. As a result, the original dye molecules tend to collect or aggregate, forming new molecules called *dimers* having the same chemical composition but of modified shape and structure. Therefore, these aggregates exhibit different energy levels, resulting in changes in absorption and emission characteristics. Unfortunately, dimers will not lase. This problem can be solved by the addition of a small quantity of de-aggregative or anti-dimerization agent (sometimes called a surface-active agent or surfactant) to the dye solution. A number of commercial detergents have been used as effective anti-dimerization agents, including liquid dishwashing preparations. In certain instances surfactants have the appropriate energy-level structure to act as triplet-quenching agents as well as to prevent the formation of dimers. Strictly speaking, dimers are aggregates of two molecules, while "trimers" are combinations of three molecules, and so on. The general term that describes a large number of molecules in an aggregate is polymers.

Tuning Mechanisms

CW dye lasers may be tuned to any wavelength over wide ranges for each dye used. Tuning may be accomplished by several methods. The most common methods will be described here.

Figure is a diagram of a birefringent tuning element employed in many CW dye lasers. It consists of three birefringent elements mounted together at Brewster's angle. Light traveling through these elements is resolved into two components, one polarized along the fast axis and one polarized along the slow axis. These two components travel at different speeds and, thus, become more out-of-phase as they travel through the component. The first element of the birefringent filter is cut to a thickness that results in a retardation of one full wavelength in the visible region for the slow ray. This element is called a full-wave plate. When white light travels through it, one wavelength will actually be retarded by exactly one wavelength. Other wavelengths will be retarded slightly more or less. The wavelength that is retarded by exactly one full wavelength will emerge with its polarization unchanged. All other wavelengths will have an elliptical polarization with a horizontal component. These horizontal components will be reflected from Brewster's-angle surfaces in the system, producing losses for all wavelengths except the one passed unchanged by the filter.

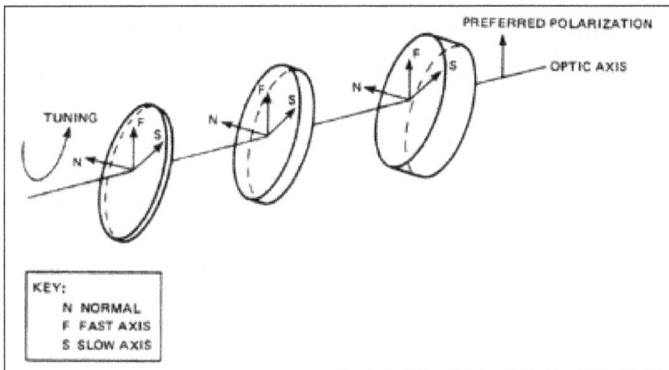

Three-element birefringent filter.

Additional filter elements are added for narrower bandwidths. The second element is twice the thickness of the first, and the third element is four times the thickness of the first. Each additional element further reduces the output linewidth. The three-element filter shown in figure has a typical output bandwidth of 0.025 nm.

The birefringent filter is tuned by rotation about an axis perpendicular to its optical surfaces. If the filter is positioned so that its "slow" axis is horizontal, the slow component of the light experiences the greatest retardation. As the filter is rotated, the angle between the slow axis and the direction of travel of the light changes becoming a minimum when the slow axis lies in a vertical plane. Reduction of this angle also reduces the retardation effect. This allows the slow ray to travel faster as the slow axis becomes more vertical. The result is that different wavelengths experience exactly one full wave retardation at different angular orientations of the filter.

A second method of tuning CW dye lasers is the tuning wedge shown in figure. This element is essentially a thin wedged etalon placed in the optical cavity of the dye laser. Tuning is accomplished by moving the tuning wedge back and forth in the cavity to allow the beam to pass through different thicknesses of the etalon.

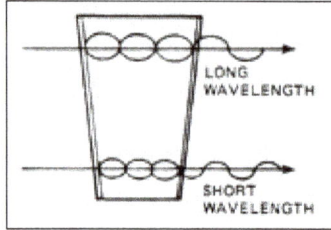

Operation of a tuning wedge.

At any particular point on the etalon, light can pass through it with low loss only if the wavelength of light is correct for forming a standing wave inside the etalon cavity. All other wavelengths experience higher loss due to the etalon and will not lase. The etalon has only a slight wedge, and sliding it through the cavity will tune the laser over a range slightly greater than the visible spectrum.

The tuning wedge is composed of a fused quartz substrate. A reflective coating is deposited on this substrate to form the first partially-reflective surface of the etalon. The next coating element is a wedge-shaped coating of transparent material that forms the wedge-shaped spacer of the etalon. The final coating is another partially-reflective coating. The tuning wedge is mounted at Brewster's angle to minimize reflection losses.

A CW dye laser with either a birefringent filter or a tuning wedge will have several longitudinal cavity modes present in the laser output. For single-mode operation, a thicker etalon for mode selection must be added to the cavity. In some cases, two etalons are added for greater control of the exact laser frequency.

Figure shows a ring dye laser designed for what is referred to as "single-frequency" operation. This laser produces an output with a spectral bandwidth of about 5 MHz.

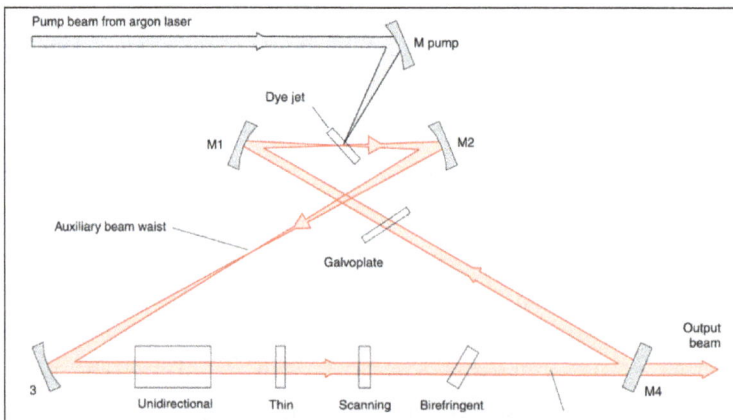

Single-frequency ring dye laser.

A ring laser has two advantages over one with a standing wave inside the optical cavity. First, the ring laser has no modes formed by the overall optical cavity and can, thus, lase at any frequency within the laser gain curve. Second, the ring laser makes greater use of the gain volume of the active medium. Standing waves in a typical laser cavity have nodes at which the intensity of the electromagnetic field of the laser light is always zero. No stimulated emission can occur at these points. This results in significant un-used gain in the active medium. In a ring laser there is only a traveling wave moving in one direction. This produces a uniform stimulation inside the active medium resulting in the extraction of more of the stored energy.

In a typical laser cavity a standing wave is formed by two waves traveling in oppo-site directions in the cavity. In the ring laser one of these waves is eliminated by an unidirectional device composed of a Faraday rotator and polarizer. This device allows polarized light to pass through in one direction only. The other elements in the optical cavity of the laser in figure commonly are found in the cavities of both ring dye lasers and those with conventional cavities. The scanning etalon is added to produce a scan of the output laser wavelength over a specified spectral range. The galvoplate is used to change the laser wavelength very slightly by tilting to change the optical length of the cavity slightly.

Extremely short laser pulses may be produced with dye lasers by using a mode-locked argon laser as a pumping source and making the dye laser cavity the same length as the argon cavity. This results in synchronous pumping in which the active medium is pumped by a mode-locked argon laser pulse at precisely the right time to produce the maximum gain for a pulse in the dye laser cavity.

A variety of laser dyes are available for use throughout the visible spectrum and into the edges of the infrared and ultraviolet regions. Figure shows the output powers and wavelengths of several of the most common dyes used in CW dye lasers.

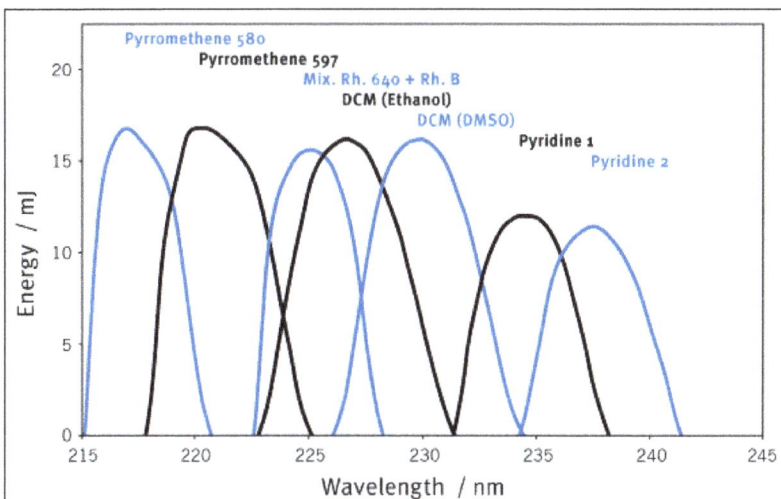

Dye laser output curves of some common laser dyes.

Nitrogen-pumped Dye Lasers

Another optical source used for optical pumping of dye lasers is the pulsed nitrogen laser. This discusses the basic characteristics of nitrogen lasers and the dye lasers used with them.

Nitrogen Lasers

Basically, nitrogen lasers are superradiant electrical discharges capable of producing up to several megawatts peak output at high repetition rates (10-500 Hz), with short output pulses of 10-20 nsec duration.

A number of different geometries for N_2 laser resonators have been employed. A laser channel (~1 meter long) of narrow, rectangular cross section is shown in Figure.

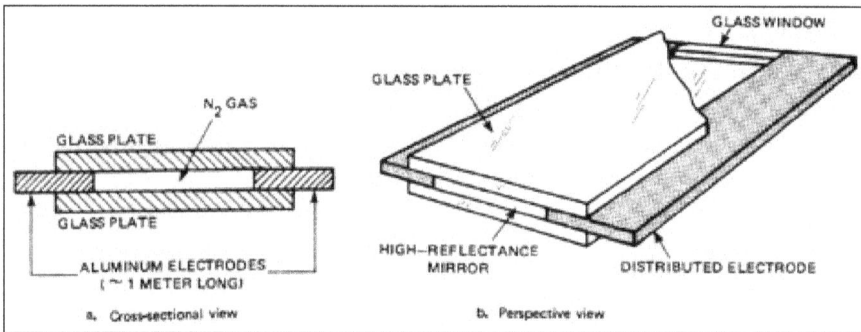

Flat N_2 laser channel.

Nitrogen lasers may be either flowing gas systems, with the gas circulated through the channel by a mechanical pump, or sealed-off systems. Both transverse and longitudinal excitation have been used. The channel illustrated in figure is excited by electrical discharges transverse to the direction of coherent light output, i.e., the optical axis. Glass plates form the top and bottom of the channel. Long electrodes, sometimes called distributed electrodes, are sandwiched between the plates to form the sides of the resonator. Although nitrogen lasers will operate without cavity mirrors (superradiant laser emission), use of a high-reflectance rear mirror increases the peak power output, decreases beam divergence, and improves beam homogeneity. The ultraviolet output of the laser exits through the glass output window. The beam, like the channel itself, is of rectangular cross section.

The energy storage/discharge circuit of an N_2 laser generally consists of five basic elements:

1. External high-voltage power supply.

2. Capacitor bank.

3. Fast switch (hydrogen thyratron).

4. Impedance-matching network (transmission network).

5. Laser channel.

Energy is stored in a capacitor bank charged by the external power supply. A trigger pulse then is applied to the grid of a hydrogen thyratron used for switching. When the thyratron conducts, the high-voltage pulse travels through the transmission line and appears across the electrodes of the laser channel. This results in a rapid discharge that excites most of the nitrogen molecules in the active medium. The excited nitrogen has an extremely high gain and lases for a few nanoseconds. This produces a high-peak-power pulse, but quickly destroys the population inversion.

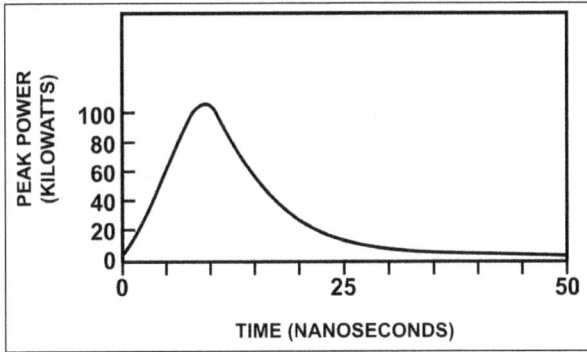

Output pulse of a nitrogen laser.

Figure illustrates a typical nitrogen laser pulse. The spectral output of N_2 lasers is concentrated in a number of lines centered around 337.1 nm in the ultraviolet. This is especially convenient for pumping a large number of organic dyes. Nitrogen-pumped dye lasers provide a powerful, tunable, coherent light source for many spectroscopic applications.

Dye Lasers for Nitrogen Pumping

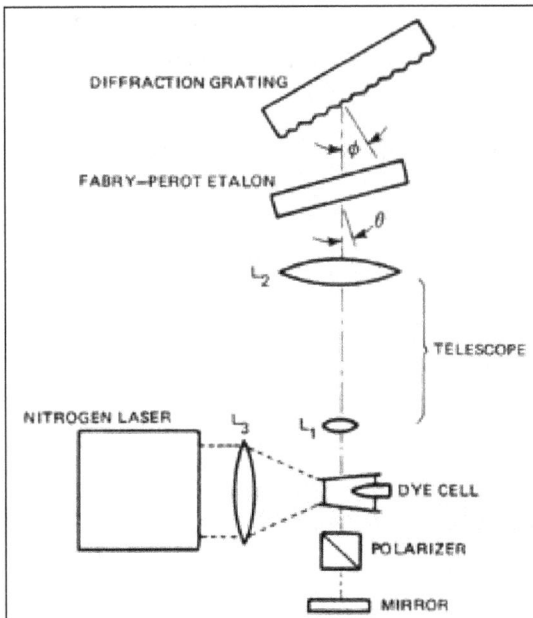

N_2-laser-pumped dye laser.

The basic layout of a tunable dye laser side-pumped by a nitrogen laser is shown in figure. Wavelength selection is provided by the diffraction grating, which also acts as the rear (HR) mirror in the cavity. The grating is mounted in a precision laser mount, capable of 0.1 arc-sec resolution. Resolution of the diffraction grating is about 0.008 nm. Addition of the Fabry-Perot etalon increases the resolution such that the resultant laser bandwidth is greatly narrowed down to about 0.0004 nm. Lens L_3 focuses the output of the N_2 laser onto the dry cell. Lenses L_1 and L_2 form an inverted astronomical telescope. The telescope expands and collimates the beam emitted by the dye solution to illuminate a large portion of the grating area for good resolution and to prevent the formation of hot spots and resultant damage on the surface of the grating.

For organic dyes with high gain, a flat quartz window is often used as the output coupler. A broadband multilayer dielectric coated mirror of about 30-50% transmittance is used for dyes having lower gain.

Flashlamp-pumped Dye Lasers

Flashlamp-pumped dye laser systems can excite a larger number of dyes, with higher peak power outputs, and they are of simpler construction, in general, than CW dye lasers. Several types of flashlamp-pumped dye lasers have been devised, differing primarily in flashlamp construction, cavity geometry, and to a lesser extent, in discharge circuitry.

Basic Construction and Operation

Figure shows the dye cell of one type of commercially available flashlamp-pumped dye laser. This is a coaxial flashlamp-dye tube design in which the flashlamp is an annular ring surrounding the dye cell for even illumination. Other pulsed dye lasers employ linear flashlamps mounted beside linear dye tubes inside elliptical pumping cavities similar to those used in solid-state lasers.

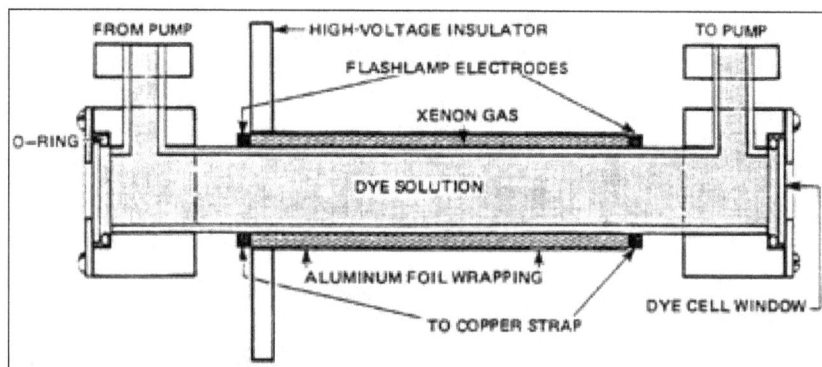

Pulsed dye laser with coaxial dye tube and flashlamp.

The flashlamp must have a short pulse for efficient pumping of the laser dye. This is usually accomplished by operation of the lamp at high voltages and currents. The energy for lamp operation is stored in a high-voltage capacitor. The capacitor is connected to the lamp, and the lamp is connected to ground through a triggered spark gap. Figure shows such a spark gap. It contains three electrodes in a pressurized nitrogen atmosphere. A standard spark plug forms two of these electrodes and is operated by an ignition coil. The third electrode is connected to the lamp terminal. The nitrogen prevents electrical conduction until the gap is triggered.

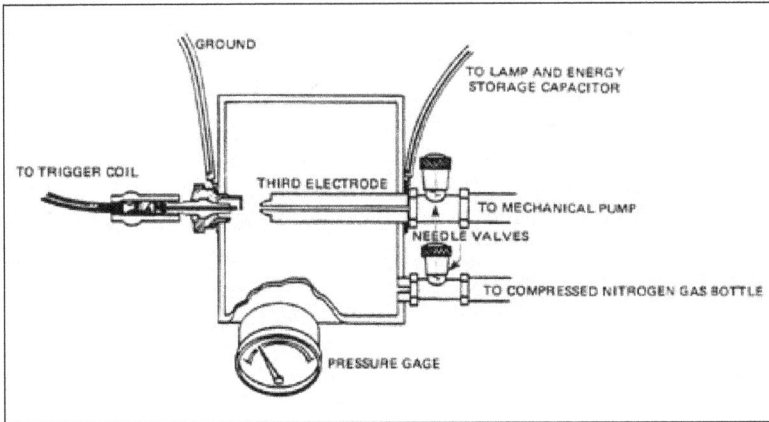

Spark gap assembly of pulsed dye laser.

The basic operation of the laser is as follows: The capacitor is charged by an external, high-voltage (0-30 kV) d.c. power supply. A trigger pulse is applied to the primary of an auto ignition coil (pulse transformer). A large voltage pulse (10-25 kV) is developed across the secondary terminals of the coil and applied to the trigger portion of the pressurized spark gap, which results in voltage breakdown of the nitrogen gas inside the chamber. This allows the charged capacitor to discharge through the spark gap and, hence, through the flashlamp.

Factors affecting Pulse Duration

In the absence of triplet-quenching (TQ) agents, a dye laser pulse would terminate in about 0.1 μ sec. Hence, much attention has been paid to fabrication of flashlamps with fast risetimes. This is reflected in the coaxial flashlamp design shown in Figure, which has a risetime of about 300-400 nsec, if a low-inductance (< 20-30 nH) discharge capacitor is used and electrical leads are kept short. The use of COT with rhodamine 6G dye solutions, and other TQ agents for various organic dyes, has led to a relaxation of the requirements for fast-risetime flashlamps. Conventional xenon-filled linear flashlamps with much slower risetimes have been used successfully for dye laser excitation.

Nevertheless, a persistent problem associated with flashlamp-pumped dye lasers is the early termination of the laser pulse, usually before the pump pulse has fallen below the level normally required for threshold excitation. Early termination was first attributed to triplet absorption and, in fact, closer synchronization of pump and laser pulses was achieved through the use of heavy concentrations of TQ agents and/or filtering out the UV components in the special output of the flashlamp(s) used. However, in the short-pulse (1-5 μ sec) regime, even with the use of TQ agents, fast discharge circuitry, and fast risetime flashlamps, shock waves (with their concomitant pressure variations in the dye solutions and resultant localized changes in the refractive index of the solution) may be a contributing factor in early termination of lasing. Finally, if water is interposed between the flash lamp and dye cell, the remaining lack of synchronization between pump and laser pulses virtually disappears, presumably because of absorption of the IR components of the flashlamp output by the water.

Wavelength Selection in Pulsed Dye Lasers

The same methods used for wavelength selection and tuning in CW dye lasers may be used with pulsed dye lasers. Diffraction gratings also may be used as with nitrogen-pumped dye lasers. A light beam incident upon a plane reflection grating will be reflected back along the axis of incidence if the following grating equation is satisfied:

$$m \lambda = 2 d \sin \theta$$

m = Diffraction order.

λ = Wavelength.

where,

d = Grating spacing.

θ = Angle of incidence measured from the normal to the grating.

A diffraction grating used in this manner is said to be mounted in the Littrow configuration. The lasing wavelength can thus be varied by rotating the grating about an axis parallel to its grooves since this configuration acts as a tunable reflector. It has high reflectivity at one wavelength at a time, depending upon its angular setting.

The metallic coatings used in gratings for commercial dye lasers are usually made of aluminum, gold, or enhanced silver for operation between 400-1100 nm. Such coatings are very fragile. Thus, a beam-expanding telescope is sometimes used between the dye cell and the grating to illuminate most of the surface of the grating to prevent the formation of "hot spots".

Advantages of Dye Lasers

- It is available in visible form (also in non-visible).
- Range of wavelengths can be produced by the using dye lasers.
- Beam diameter is very less.
- Its beam divergence (0.8 milli radians to 2 milli radians) is also less from many lasers beam divergence.

- Construction of dye laser is not so complex.

- Having the greater efficiency 25%.

- High output power is also possible with dye lasers.

Disadvantages of Dye Lasers

- Cost of dye lasers is very high.

- Some cases need other laser beam.

- To tune at one frequency, the laser uses birefringent element or filter making it more costly.

- In dye lasers it is very difficult to determine the element that actually lases because dye has complex chemical formula.

Semiconductor Laser

Semiconductor lasers are lasers based on semiconductor gain media, where optical gain is usually achieved by stimulated emission at an interband transition under conditions of a high carrier density in the conduction band.

Design Features

In reality a semiconductor laser is simply a semiconductor diode, because its active medium is the junction of the forward biased P-N diode, shown as:

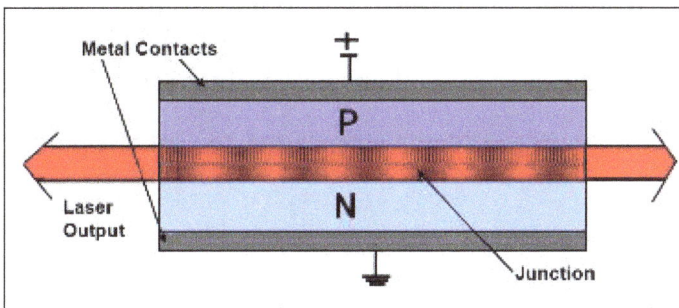

Here the metal contacts shown are used to connect the P-N material to the DC power supply. The junction shown is few micrometers thick. At the junction light is emitted when electrons or current pass from N to P type material. In other words, current is injected into the junction between N and P type materials. It is why we use to semiconductor laser the name of Injection Laser also. Since we know that a minimum current density (similar to the gain threshold) is necessary for the occurrence of lasing. So when the minimum current density is reached then increasing the current density across the junction region will increase the output of the laser.

Unlike other lasers, semiconductor laser does not need mirrors to obtain the reflectivity needed to produce feedback mechanism. Reflection from the cleaved ends of the semiconductor chip is enough to produce lasing.

The reflectivity of the interface between the semiconductor material and air is approximately 36% which is enough to provide adequate feedback as well as serve as the output coupler. The beam although exist from both ends. If a beam is desired from one end of the laser only then opposite ends can be coated to reflect higher amounts of light.

The temperature has great effect on the output of the semiconductor laser. When the temperature increases then significant power losses occurs within the laser. That is why the semiconductor laser is sometimes cooled by liquid nitrogen or some other cooling system. However these lasers can be operated at room temperature if the losses are acceptable and current density is high enough.

Pulsing the current leads us to very much improved performance. Gallium Arsenide lasers are usually operated in pulsed mode, with duty cycle less than 1% because higher duty cycles cause an increased temperature, which greatly affect the output characteristics. The gallium arsenide laser produces light in near infrared spectrum ranging from 845nm to 905nm. The lasing medium of the semiconductor laser is short and rectangular. Therefore the output beam does not have the same dimension in both vertical and horizontal axis. Hence the beam profile has an unusual shape. The beam divergence of semiconductor lasers is much greater than most of the lasers, depending on temperature, therefore ranging from 125 to 400 milli radians.

In spite of the fact that semiconductor lasers do not produce a beam with characteristics as desired in other types of lasers, their small size, low power consumption and relatively high ef high efficiency make them very useful device.

Common materials for semiconductor lasers (and for other optoelectronic devices) are:

- GaAs (gallium arsenide),
- AlGaAs (aluminum gallium arsenide),
- GaP (gallium phosphide),
- InGaP (indium gallium phosphide),
- GaN (gallium nitride),
- InGaAs (indium gallium arsenide),

- GaInNAs (indium gallium arsenide nitride),

- InP (indium phosphide),

- GaInP (gallium indium phosphide).

These are all direct bandgap semiconductors; indirect bandgap semiconductors such as silicon do not exhibit strong and efficient light emission. As the photon energy of a laser diode is close to the bandgap energy, compositions with different bandgap energies allow for different emission wavelengths. For the ternary and quaternary semiconductor compounds, the bandgap energy can be continuously varied in some substantial range. In AlGaAs = $Al_xGa_{1-x}As$, for example, an increased aluminum content (increased x) causes an increase in the bandgap energy.

While the most common semiconductor lasers are operating in the near-infrared spectral region, some others generate red light (e.g. in GaInP-based laser pointers) or blue or violet light (with gallium nitrides). For mid-infrared emission, there are e.g. lead selenide (PbSe) lasers (lead salt lasers) and quantum cascade lasers.

Apart from the above-mentioned inorganic semiconductors, organic semiconductor compounds might also be used for semiconductor lasers. The corresponding technology is by far not mature, but its development is pursued because of the attractive prospect of finding a way for cheap mass production of such lasers. So far, only optically pumped organic semiconductor lasers have been demonstrated, whereas for various reasons it is difficult to achieve a high efficiency with electrical pumping.

Typical Characteristics and Applications

Some typical aspects of semiconductor lasers are:

- Electrical pumping with moderate voltages and high efficiency is possible particularly for high-power diode lasers, and allows their use e.g. as pump sources for highly efficient solid-state lasers (→ diode-pumped lasers).

- A wide range of wavelengths are accessible with different devices, covering much of the visible, near-infrared and mid-infrared spectral region. Some devices also allow for wavelength tuning.

- Small laser diodes allow fast switching and modulation of the optical power, allowing their use e.g. in transmitters of fiber-optic links.

Such characteristics have made semiconductor lasers the technologically most important type of lasers. Their applications are extremely widespread, including areas as diverse as optical data transmission, optical data storage, metrology, laser spectroscopy, laser material processing, pumping solid-state lasers (→ diode-pumped lasers), and various kinds of medical treatments.

Pulsed Output

Most semiconductor lasers generate a continuous output. Due to their very limited energy storage capability (low upper-state lifetime), they are not very suitable for pulse generation with Q switching, but quasi-continuous-wave operation often allows for significantly enhanced powers. Also, semiconductor lasers can be used for the generation of ultrashort pulses with mode locking or gain switching. The average output powers in short pulses are usually limited to at most a few milliwatts, except for optically pumped surface-emitting external-cavity semiconductor lasers (VECSELs), which can generate multi-watt average output powers in picosecond pulses with multi-gigahertz repetition rates.

Modulation and Stabilization

A particular advantage of the short upper-state lifetime is the capability of semiconductor lasers to be modulated with very high frequencies, which can be tens of gigahertz for VCSELs. This is exploited mainly in optical data transmission, but also in spectroscopy, for the stabilization of lasers to reference cavities, etc.

Advantages of Semiconductor Lasers

- Smaller size and appearance make them good choice for many applications.
- From cost point of view the semiconductor lasers are economical.
- Semiconductor lasers construction is very simple.
- No need of mirrors is in semiconductor lasers.
- Semiconductor lasers have high efficiency.
- The low power consumption is also its great advantage.

Disadvantages of Semiconductor Lasers

- Due to relatively low power production, these lasers are not suited to many typical laser applications.
- Semiconductor laser is greatly dependent on temperature.
- The temperature affects greatly the output of the laser.
- The lasing medium of semiconductor lasers is too short and rectangular so the output beam profile has an unusual shape.
- Beam divergence is much greater from 125 to 400 milli radians as compared to all other lasers.
- The cooling system requirement in some cases may be considered its disadvantage.

Laser Diode

A laser diode is a semiconductor laser that is closely related to the light emitting diode (LED) both in form and in operation. The laser diode is quite different from the common perception of lasers as big, bulky and power-hungry devices that emit an intense beam of light that can burn or even cut. A laser diode can be considered as an LED that emits focused light and is commonly used in everyday consumer devices such as DVD and Blu-ray players, barcode scanners, fiber optic communication equipment and laser printers.

The laser diode was invented by Dr. Robert Hall of General Electric. He filed his patent on October 24, 1962, and received the patent grant on April 5, 1966, as US Patent no. 3,245,002. The diode emits laser light by bouncing photons back and forth between slices of p-type and n-type semiconductors that are roughly 1 micrometer apart. This is a similar process used in conventional lasers that produce beams by repeatedly pumping the light emitted by the atoms between two mirrors.

The forward biasing done on the semiconductors used in the laser diode forces the two charge carriers (holes and electrons) to be injected or pumped from the opposite side of the p-n junction into the other side called the depletion region, which has zero charge carriers. Because of this, laser diodes are also called injection laser diodes.

Laser Diode Construction

The above figure shows a simplified construction of a laser diode, which is similar to a light emitting diode (LED). It uses gallium arsenide doped with elements such as selenium, aluminium, or silicon to produce P type and N type semiconductor materials. While a laser diode has an additional active layer of undoped (intrinsic) gallium arsenide have the thickness only a few nanometers, sandwiched between the P and N layers, effectively creating a PIN diode (P type-Intrinsic-N type). It is in this layer that the laser light is produced.

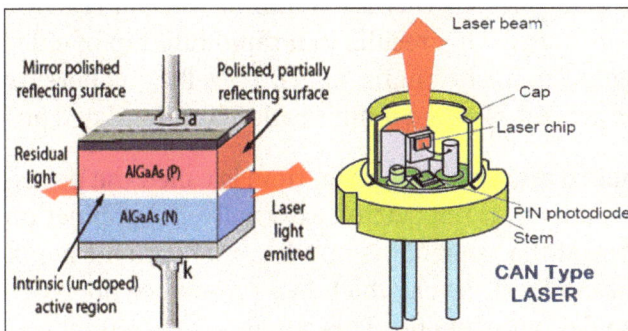

Laser Diode Construction.

Working Principles of Laser Diode

Every atom according to the quantum theory, can energies only within a certain discrete energy level. Normally, the atoms are in the lowest energy state or ground state. When an energy source given to the atoms in the ground state can be excited to go to one of the higher levels. This process is called absorption. After staying at that level for a very short duration, the atom returns to its initial ground state, emitting a photon in the process. This process is called spontaneous emission. These two processes, absorption and spontaneous emission, take place in a conventional light source.

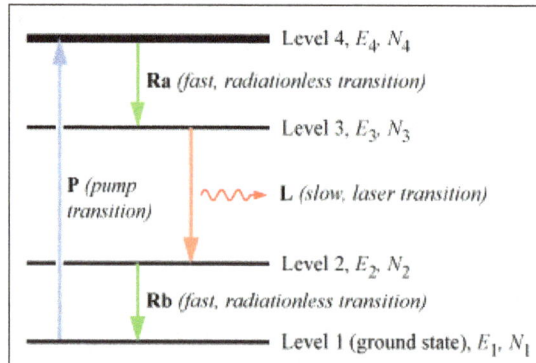

Principle of Laser Action.

In case the atom, still in an excited state, is struck by an outside photon having precisely the energy necessary for spontaneous emission, the outside photon is increased by the one given up by the excited atom. Moreover, both the photons are released from the same excited state in the same phase. This process, called stimulated emission, is fundamental for laser action. In this process, the key is the photon having exactly the same wavelength as that of the light to be emitted.

Amplification and Population Inversion

When favourable conditions are created for the stimulated emission, more and more atoms are forced to emit photons thereby initiating a chain reaction and releasing an enormous amount of energy. This results in a rapid buildup of energy of emitting one particular wavelength (monochromatic light), travelling coherently in a particular, fixed direction. This process is called amplification by stimulated emission.

The number of atoms in any level at a given time is called the population of that level. Normally, when the material is not excited externally, the population of the lower level or ground state is greater than that of the upper level. When the population of the upper level exceeds that of the lower level, which is a reversal of the normal occupancy, the process is called population inversion. This situation is essential for a laser action. For any stimulated emission.

It is necessary that the upper energy level or met stable state should have a long life-time, i.e., the atoms should pause at the met stable state for more time than at the lower level. Thus, for laser action, pumping mechanism (exciting with external source) should be from as such, as to maintain a higher population of atoms in the upper energy level relative to that in the lower level.

It is necessary that the upper energy level or met stable state should have a long lifetime, i.e., the atoms should pause at the met stable state for more time than at the lower level. Thus, for laser action, pumping mechanism (exciting with external source) should be from as such, as to maintain a higher population of atoms in the upper energy level relative to that in the lower level.

Controlling the Laser Diode

The laser diode is operated at a much higher current, typically about 10 times greater than a normal LED. The below figure compares a graph of the light output of a normal LED and that of a laser diode. In a LED the light output increases steadily as the diode current is increased. In a laser diode, however laser light is not produced until the current level reaches the threshold level when stimulated emission starts to occur. The threshold current is normally more than 80% of the maximum current the device will pass before being destroyed. For this reason, the current through the laser diode must be carefully regulated.

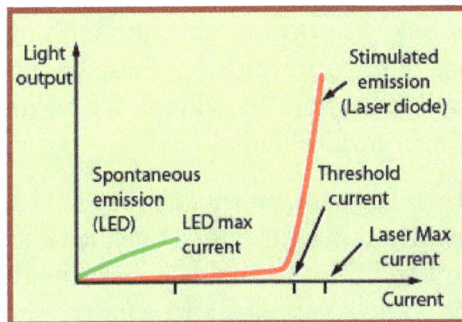

Comparison between a LED.

Another problem is that the emission of photons is very dependent on temperature the diode is already being operated close to its limit and so gets hot, therefore changing the amount of light emitted (photons) and the diode current. By the time the laser diode is working efficiently it is operating on the brink of disaster. If the current reduces and falls below the threshold current, stimulated emission ceases; just a little too much current and the diode is destroyed.

As the active layer is filled with oscillating photons, some (typically about 60%) of the light escapes in a narrow, flat beam from the edge of the diode chip. As shown in figure, some residual light also escapes at the opposite edge and is used to activate a photodiode, which converts the light back into the electric current. This current is used as a feedback to the automatic diode driver circuit, to measure the activity in the laser diode and so make sure by controlling the current through the laser diode, that the current and light output remain at a constant and safe level.

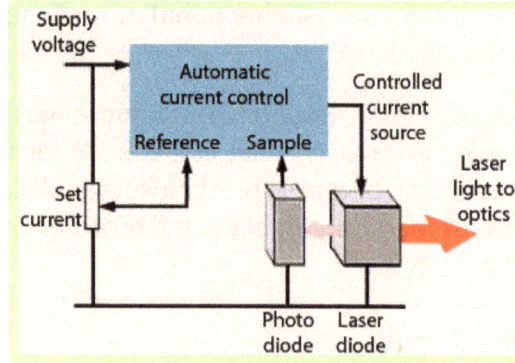

Controlling the Laser Diode.

Applications of Laser Diode

Laser Diode Modules are ideal for applications such as life science, industrial, or scientific instrumentation. Laser Diode Modules are available in a wide variety of wavelengths, output powers, or beam shapes. Low power Lasers are used in an increasing number of familiar applications including CD and DVD players and recorders, bar code readers, security systems, optical communications and surgical instruments:

1. Industrial applications: Engraving, cutting, scribing, drilling, welding, etc. Medical applications remove unwanted tissues, diagnostics of cancer cells using fluorescence, dental medication. In general, the results using lasers are better than the results using a surgical knife.

2. Laser Diodes used for Telecom: In the telecom field 1.3 μm and 1.55 μm band laser diodes used as the main light source for silica fibre lasers have a less transmission loss in the band. The laser diode with the different band is used for pumping source for optical amplification or for the short-distance optical link.

Types of Laser Diodes

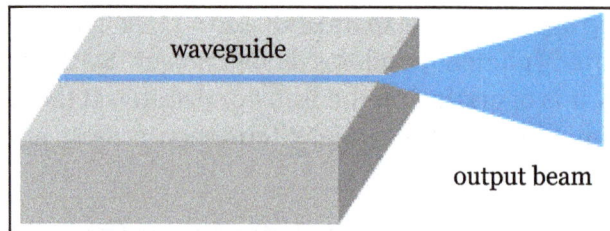

Schematic setup of an edge-emitting low-power laser diode.
The waveguide and the output beam emerging at one edge of the
wafer die are shown, but not the electrode structures.

Most laser diodes (LDs) are built as edge-emitting lasers, where the laser resonator is formed by coated or uncoated end facets (cleaved edges) of the semiconductor wafer. They are often based on a double heterostructure, which restricts the generated carriers to a narrow region and at the same time serves as a waveguide for the optical field (double confinement). The current flow is restricted to the same region, sometimes using isolating barriers. Such arrangements lead to a relatively low threshold pump power and high efficiency. The active region is usually quite thin – often so thin that it acts as a quantum well.

Some modern kinds of LDs are of the surface-emitting type where the emission direction is perpendicular to the wafer surface, and the gain is provided by multiple quantum wells.

There are very different kinds of LDs, operating in very different regimes of optical output power, wavelength, bandwidth, and other properties:

- Small edge-emitting LDs generate between a few milliwatts and up to roughly half a watt of output power in a beam with high beam quality. The output may be emitted into free space or coupled into a single-mode fiber. Such lasers can be designed to be either *index guiding* (with a waveguide structure guiding the laser light within the LD) or gain guiding (where the beam profile is kept narrow via preferential amplification on the beam axis).

- Small LDs made as distributed feedback lasers (DFB lasers) or distributed Bragg reflector lasers (DBR lasers) with short resonators can achieve single-frequency operation, sometimes combined with wavelength tunability.

- External cavity diode lasers contain a laser diode as the gain medium of a longer laser resonator, completed with additional optical elements such as laser mirrors or a diffraction grating. They are often wavelength-tunable and exhibit a small emission linewidth.

- Broad area laser diodes (also often called broad stripe laser diodes or wide stripe lasers) generate up to a few watts of output power. The beam quality is significantly poorer than that of lower-power LDs, but better than that of diode bars.

- High brightness laser diodes are laser diodes which are optimized for a particularly high radiance (brightness). Different technologies may be used, and such lasers are available on quite different power levels.

- Slab-coupled optical waveguide lasers (SCOWLs), containing a multi-quantum well gain region in a relatively large waveguide, can generate a watt-level output in a diffraction-limited beam with a nearly circular profile.

- High-power diode bars contain an array of broad-area emitters, generating tens of watts with poor beam quality. Despite the higher power, the brightness is lower than that of a broad area LD.

- High-power stacked diode bars (→ diode stacks) are stacks of multiple diode bars for the generation of extremely high powers of hundreds or thousands of watts.

- Monolithic surface-emitting semiconductor lasers (VCSELs) typically generate a few milliwatts with high beam quality. There are also external-cavity versions of such lasers (VECSELs) which can generate much higher powers with still excellent beam quality.

Laser diodes may emit a beam into free space, but many LDs are also available in fiber-coupled form. The latter makes it particularly convenient to use them, e.g., as pump sources for fiber lasers and fiber amplifiers.

Vertical-cavity Surface Emitting Laser

Vertical-Cavity Surface-Emitting Lasers (VCSELs) are a relatively recent type of semiconductor lasers. VCSELs were first invented in the mid-1980's. Very soon, VCSELs gained a reputation as a superior technology for short reach applications such as fiber channel, Ethernet and intra-systems links. Then, within the first two years of commercial availability, VCSELs became the technology of choice for short range datacom and local area networks, effectively displacing edge-emitter lasers. This success was mainly due to the VCSEL's lower manufacturing costs and higher reliability compared to edge-emitters.

Princeton Optronics has developed the key technologies resulting in the world's highest power single VCSEL devices and 2-D arrays. We have successfully demonstrated single devices with >5W CW output power and large 2D arrays with >230W CW output power. We have made single mode devices of 1W output power and single mode arrays with power of >100W which are coupled to 100u, 0.22NA fiber. The highest wall plug efficiency of these devices and arrays is 56%. We have made arrays which deliver 1kW/cm^2 in CW operation and 4.2kW/cm^2 in QCW operation. Princeton Optronics was a participant in the DARPA-SHEDS program, whose main objective was to improve laser diode conversion efficiency.

VCSEL Structure

Semiconductor lasers consist of layers of semiconductor material grown on top of each other on a substrate (the "epi"). For VCSELs and edge-emitters, this growth is typically done in a molecular-beam-epitaxy (MBE) or metalorganic-chemical-vapor-deposition (MOCVD) growth reactor. The grown wafer is then processed accordingly to produce individual devices. Figure summarizes the differences between VCSEL and edge-emitter processing.

In a VCSEL, the active layer is sandwiched between two highly reflective mirrors (dubbed distributed Bragg reflectors, or DBRs) made up of several quarter-wavelength-thick layers of semiconductors of alternating high and low refractive index. The reflectivity of these mirrors is typically in the range 99.5 ~99.9%. As a result, the light oscillates perpendicular to the layers and escapes through the top (or bottom) of the device. Current and/or optical confinement is typically achieved through either selective-oxidation of an Aluminum-rich layer, ion-implantation, or even both for certain applications. The VCSELs can be designed for "top-emission" (at the epi/air interface) or "bottom-emission" (through the transparent substrate) in cases where "junction-down" soldering is required for more efficient heat-sinking for example. Figure illustrates different common types of VCSEL structures.

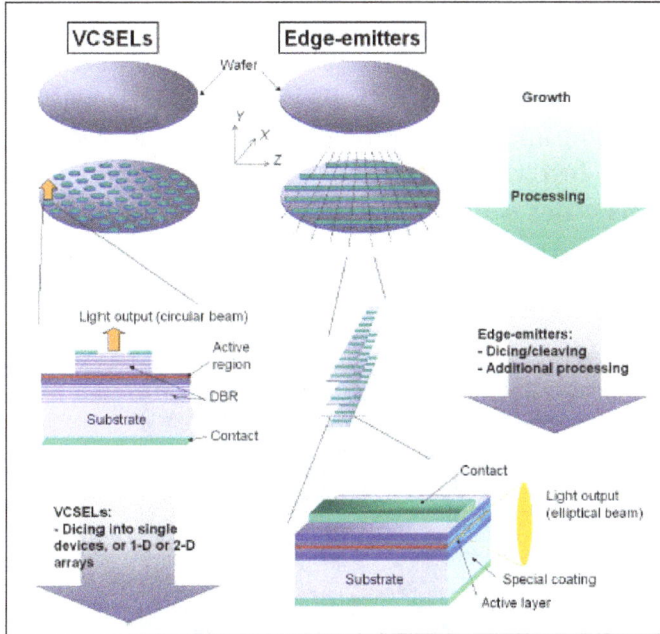

Comparison of the growth/processing flow
of VCSEL and edge-emitter semiconductor lasers.

In contrast, edge-emitters are made up of cleaved bars diced from the wafers. Because of the high index of refraction contrast between air and the semiconductor material, the two cleaved facets act as mirrors. Hence, in the case of an edge-emitter, the light oscillates parallel to the layers and escapes side-ways. This simple structural difference between the VCSEL and the edge-emitter has important implications.

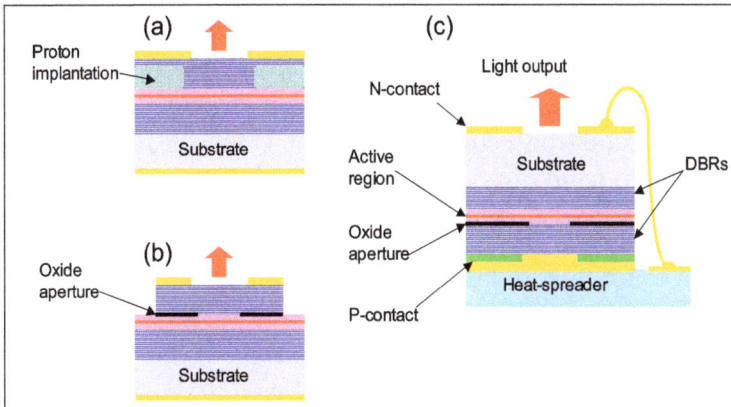

Three common types of VCSEL structures: (a) a top-emitting structure with proton implantation to confine the current, (b) a selectively-oxidized top-emitting structure to confine the optical modes and/or the current, and (c) a mounted bottom-emitting selectively-oxidized structure.

Since VCSELs are grown, processed and tested while still in the wafer form, there is significant economy of scale resulting from the ability to conduct parallel device processing, whereby equipment utilization and yields are maximized and set up times and labor content are minimized. In the case of a VCSEL, the mirrors and active region are sequentially stacked along the Y axis during epitaxial growth. The VCSEL wafer then goes through etching and metalization steps to form the electrical contacts. At this point the wafer goes to test where individual laser devices are characterized on a pass-fail basis. Finally, the wafer is diced and the lasers are binned for either higher-level assembly (typically >95%) or scrap (typically <5%). The following figure shows a single high-power VCSEL device (>2W output power) packaged on a high-thermal conductivity submount.

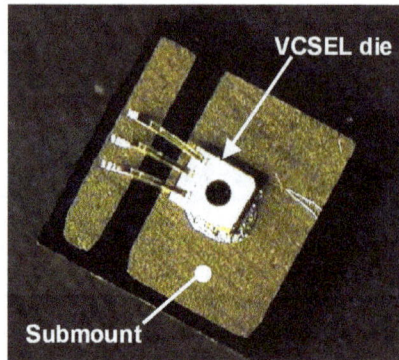

Packaged high-power VCSEL device (>2W). The submount is 2mm x 2mm.

In a simple Fabry-Pérot edge-emitter the growth process also occurs along the Y axis, but only to create the active region as mirror coatings are later applied along the Z axis. After epitaxial growth, the wafer goes through the metallization step and is subsequently cleaved along the X axis, forming a series of wafer strips. The wafer strips are then stacked and mounted into a coating fixture. The Z axis edges of the wafer strips are then coated to form the device mirrors. This coating is a critical processing step for edge-emitters, as any coating imperfection will result in early and catastrophic failure of the devices due to catastrophic-optical-damage (COD). After this coating step, the wafer strips are diced to form discrete laser chips, which are then mounted onto carriers. Finally, the laser devices go into test.

It is also important to understand that VCSELs consume less material: in the case of a 3" wafer, a laser manufacturer can build about 15,000 VCSEL devices or approximately 4,000 edge-emitters of similar power levels.

In addition to these advantages, VCSEL also demonstrate excellent dynamic performances such as low threshold currents (a few micro-amps), low noise operation and high-speed digital modulation (10 Gb/s). Furthermore, although VCSELs have been confined to low-power applications – a few milli-Watts at most – they have the inherent potential of producing very high powers by processing large 2-D arrays. In contrast, edge-emitters cannot be processed in 2-D arrays.

VCSEL Advantages

The many advantages offered by the VCSEL technology can be summarized in the following points:

1. Wavelength stability: The lasing wavelength in a VCSEL is very stable, since it is fixed by the short (1 ~1.5- wavelength thick) Fabry-Perot cavity. Contrary to edge-emitters, VCSELs can only operate in a single longitudinal mode.

2. Wavelength uniformity and spectral width: Growth technology has improved such that VCSEL 3" wafers are produced with less than a 2nm standard deviation for the cavity wavelength. This allows for the fabrication of VCSEL 2-D arrays with little wavelength variation between the elements of the array (<1nm full-width half- maximum spectral width). By contrast, edge-emitter bar-stacks suffer from significant wavelength variations from bar to bar since there is no intrinsic mechanism to stabilize the wavelength, resulting in a wide spectral width (3 ~5nm FWHM).

3. Temperature sensitivity of wavelength: The emission wavelength in VCSELs is ~5 times less sensitive to temperature variations than in edge-emitters. The reason is that in VCSELs, the lasing wavelength is defined by the optical thickness of the single-longitudinal-mode-cavity and that the temperature dependence of this optical thickness is minimal (the refractive index and physical thickness of the cavity have a weak dependence on temperature). On the other hand, the lasing wavelength in edge-emitters is defined by the peak-gain wavelength, which has a much stronger dependence on temperature. As a consequence, the spectral linewidth for high-power arrays (where heating and temperature gradients can be significant) is much narrower in VCSEL arrays than in edge-emitter-arrays (bar-stacks). Also, over a 20 °C change in temperature, the emission wavelength in a VCSEL will vary by less than 1.4nm (compared to ~7nm for edge-emitters).

4. High Temperature Operation (Chillerless operation for pumps: VCSEL devices can be operated without refrigeration- because they can be operated at temperatures to 80 °C, The cooling system becomes very small, rugged and portable with this approach.

5. Higher power per unit area: Edge emitters deliver a maximum of about 500W/ cm² because of gap between bar to bar which has to be maintained for coolant flow, while VCSELs are delivering ~1200W/ cm² now and can deliver 2-4kW/ cm² in near future.

6. Beam Quality: VCSELs emit a circular beam. Through proper cavity design VCSELs can also emit in a single transverse mode (circular Gaussian). This simple beam structure greatly reduces the complexity and cost of coupling/ beam-shaping optics (compared to edge-emitters) and increases the coupling efficiency to the fiber or pumped medium. This has been a key selling point for the VCSEL technology in low-power markets.

7. Reliability: Because VCSELs are not subject to catastrophic optical damage (COD), their reliability is much higher than for edge-emitters. Typical FIT values (failures in one billion device-hours) for VCSELs are <10.

8. Manufacturability and yield: Manufacturability of VCSELs has been a key selling point for this technology. Because of complex manufacturing processes and reliability issue related to COD (catastrophic optical damage), edge-emitters have a low yield (edge-emitter 980nm pump chip manufacturers typically only get ~500 chips out of a 2" wafer). On the other hand, yields for VCSELs exceed 90% (corresponds to ~5000 high-power chips from a 2" wafer). In fact, because of its planar attributes, VCSEL manufacturing is identical to standard IC Silicon processing.

9. Scalability: For high-power applications, a key advantage of VCSELs is that they can be directly processed into monolithic 2-D arrays, whereas this is not possible for edge-emitters (only 1-D monolithic arrays are possible). In addition, a complex and thermally inefficient mounting scheme is required to mount edge-emitter bars in stacks.

10. Packaging and heat-sinking: Mounting of large high-power VCSEL 2-D arrays in a "junction-down" configuration is straightforward (similar to micro-processor packaging), making the heat-removal process very efficient, as the heat has to traverse only a few microns of AlGaAs material. Record thermal impedances of <0.16K/W have been demonstrated for 5mm x 5mm 2-D VCSEL arrays.

11. Cost: With the simple processing and heat-sinking technology it becomes much easier to package 2-D VCSEL arrays than an equivalent edge-emitter bar-stack. The established existing silicon industry heat-sinking technology can be used for heat removal for very high power arrays. This will significantly reduce the cost of the high-power module. Currently, cost of the laser bars is the dominant cost for the DPSS lasers.

High Wavelength Stability and Low Temperature Dependence

Since the VCSEL resonant cavity is defined by a wavelength-thick cavity sandwiched between two distributed Bragg reflectors (DBRs), devices emit in a single longitudinal mode and the emission wavelength is inherently stable (<0.07nm/K), without the need for additional wavelength stabilization schemes or external optics, as is the case for edge-emitters. Furthermore, thanks to advances in growth and packaging technologies, the emission wavelength is very uniform across a 5mm x 5mm VCSEL array, resulting in spectral widths of 0.7 ~0.8nm (full-width at half- maximum).

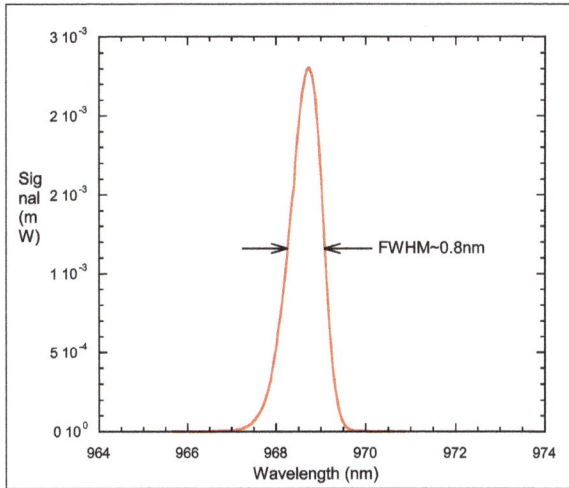

Emission spectrum of a 5mm x 5mm VCSEL array at 100W output power (120A).

This wavelength stability and narrow spectral width can be very significant advantages in pumping applications for example where the medium has a narrow absorption band.

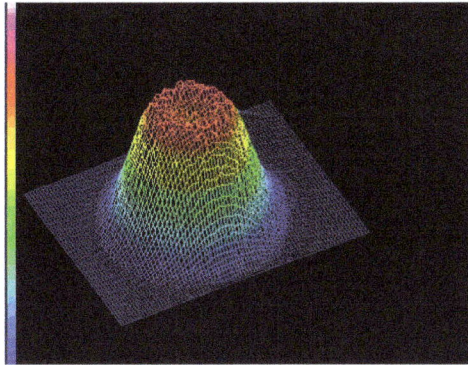

Far-field beam profile of a 5mm x 5mm VCSEL array at 100W output power (120A).

Circular Output Beam

Unlike edge-emitters, VCSELs emit in a circularly symmetric beam with low divergence without the need for additional optics. This has been a tremendous advantage for low-power VCSELs in the telecom and datacom markets because of their ability to directly couple to fibers ("butt-coupling") with high coupling efficiency. Princeton Optronics' high-power VCSEL arrays emit in a quasi-top-hat beam profile, making these devices ideal for direct pumping ("butt-pumping") of solid-state lasers.

Feedback Insensitivity

In VCSELs, the as-grown output coupler reflectivity is very high (typically >99.5%)

compared to edge-emitters (typically <5%). This makes VCSELs extremely insensitive to optical feedback effects, thus eliminating the need for expensive isolators or filters in some applications.

Low Thermal Impedance and Ease of Packaging

Princeton Optronics has developed advanced packaging technologies, which enables efficient and reliable die-attach of large 2-D VCSEL arrays on high-thermal-conductivity submounts. The resulting submodule layout allows for straightforward packaging on a heat-exchanger. For high-power devices packaged on micro-coolers, Princeton Optronics has demonstrated modules with thermal impedances as low as 0.15K/W (between the chiller and the chip active layer). Princeton Optronics can provide its customers with several heat-exchanger and heat-sinking application notes.

High Power CW and QCW VCSEL Arrays

Princeton Optronics designs and manufactures advanced high-power CW and QCW diode lasers for the industrial, medical, and defense markets. Princeton Optronics' innovative approach is based on the Vertical-Cavity Surface Emitting Laser technology (VCSEL for short), enabling us to manufacture and deliver laser diodes with exceptionally high reliability, and superior spectral and beam properties.

Arrays with Hundreds of Watts (>1kW/cm² Power Density) Output Power

Vertical-Cavity Surface-Emitting Lasers were initially introduced in the mid-90's as a low-cost alternative to edgeemitters, for use as a low-power source (sub-mW to a few mW) in datacoms and telecoms. Within two years of their introduction, VCSELs overwhelmed and replaced the edge-emitter technology in these markets due to their better beam quality, reduced manufacturing costs and much higher reliability. Now, a new class of VCSELs has been developed for high power applications. Princeton Optronics is the first company to introduce such high power VCSEL products to the market.

Unlike edge-emitters, the light emits perpendicular to wafer surface for VCSELs. It is therefore straightforward to process 2-D arrays of small VCSEL devices driven in parallel to obtain higher output powers. The advantage of 2D arrays is that it has simple silicon IC chip-like configuration and many of the silicon IC packaging and cooling technology can be applied to VCSEL arrays.

Princeton Optronics has taken the VCSEL technology to very high power levels by developing very large (5mm x 5mm) 2-D VCSEL arrays packaged on high-thermal-conductivity submounts. These arrays are composed of thousands of low-power single devices driven in parallel. Using this approach, record CW output powers in excess of 230W from a 0.22cm² emission area (>1kW/cm²) have been demonstrated, without sacrificing wall-plug efficiency.

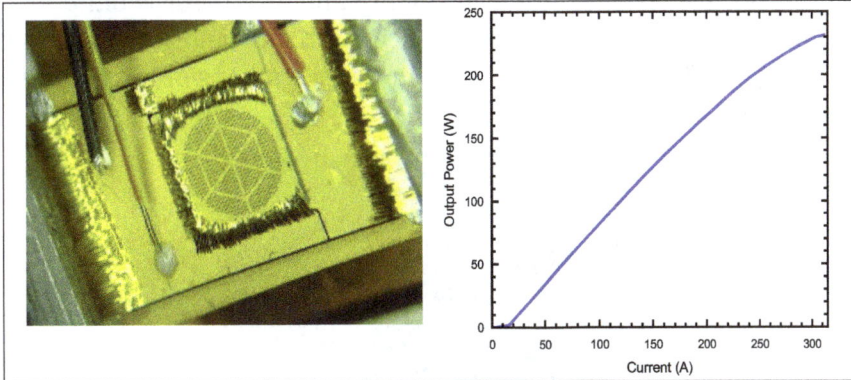

Picture of high-power 5mm x 5mm 2-D VCSEL array mounted on a micro-cooler and measure CW output power and voltage at a constant heat-sink temperature. Roll-over power is >230W.

Very High-power Density QCW Operation

In addition to CW VCSEL arrays, Princeton Optronics has developed very high power density VCSEL arrays for quasiCW (QCW) operation. QCW powers in excess of 925W have been demonstrated from very small arrays (5 x 5mm chip size), resulting in record power densities >4.2kW/cm². These small arrays can easily be connected in series to form larger arrays with high output powers.

These arrays are ideal for applications requiring very compact high-power laser sources.

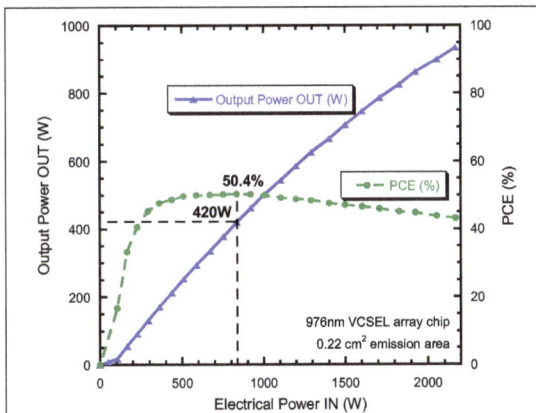

Power vs. current for a small VCSEL 2D array under different QCW regimes. These arrays exhibit power densities >4.2kW/cm².

High Temperature Operation

Because VCSELs can operate reliably at temperatures up to 80 °C, they do not necessarily require refrigeration. Additionally, since the wavelength change with temperature is small, the cooling system design can be considerably simplified. The cooling system thus becomes very small, rugged and portable with this approach. We have been operating the VCSELs and VCSEL arrays with water pump and a radiator cooling like that of a car engine. Figure shows such a set up in which a radiator and water pump is used to cool a 120W array of VCSELs. The result of the cooling arrangement is compared with a chiller cooling and shown in figure.

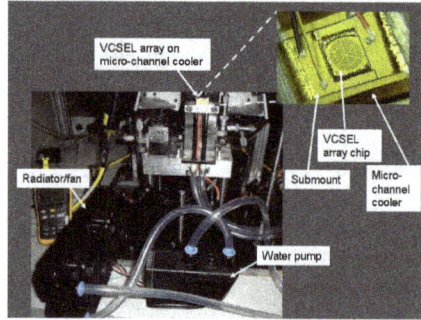

Shows the set up without chiller using a radiator and a water pump in an arrangement like in a car engine.

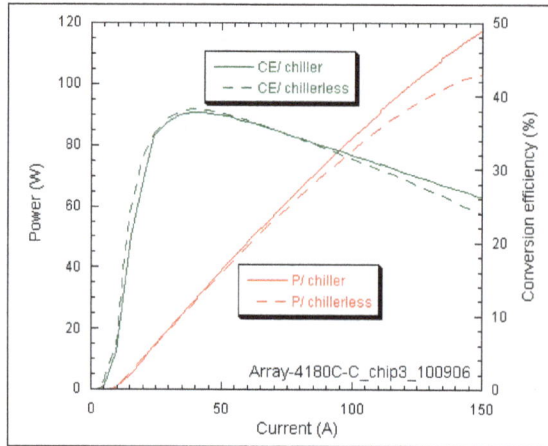

The above figure Shows the performance of a 120W VCSEL array with fan-radiator cooling with water temperature at 45 °C vs cooling with a chiller with 16 °C water temperature. The green curves show the efficiency (CE) in the two cases which is almost similar. The red curves show the power output from the array in the two cases. The power output decreases somewhat at higher power, but at power levels below 80W, there is very little change.

Gas Laser

A variety of lasers is based on gases as gain media. The laser-active entities are either single atoms or molecules, and are often used in a mixture with other substances having auxiliary functions. A population inversion as the prerequisite for gain via stimulated emission is in most cases achieved by pumping the gas with an electric discharge, but there are also gas lasers using a chemical reaction, optically pumped devices, and Raman lasers. During operation, the gas is often in the state of a plasma, containing a significant concentration of electrically charged particles.

Most gas lasers emit with a high beam quality, often close to diffraction-limited, since the gas introduces only weak optical distortions, despite considerable temperature gradients. Their operation usually requires a high-voltage supply, often with a high electrical power. Some high-power gas lasers use a system for quickly circulating the gas (forced convection, fast flow).

Laser Operation

The basic design concept of most of the gas lasers is the same. It has been schematically depicted in figure. The gaseous medium is put into a laser tube usually made of glass or some ceramic or in a chamber at a pressure in the range of a few millitorrs to several atmospheres depending on the type of laser. The gas may be flowing or be sealed in the tube. The mirrors of the optical resonators are mounted either directly at the ends of the laser tube or external to the tube. The gaseous medium is usually excited by passing an electrical discharge current through it. Some lasers are operated in a continuous wave (CW) mode, some in a pulsed mode. Continuous operation is achieved by passing either direct current or r.f. current. For pulsed operation usually a storage capacitor is charged at a high voltage and then is discharged through the gas with a fast switch normally a spark gap for single-shot operation or a thyratron for high repetition rate operation (100 Hz to 100 kHz).

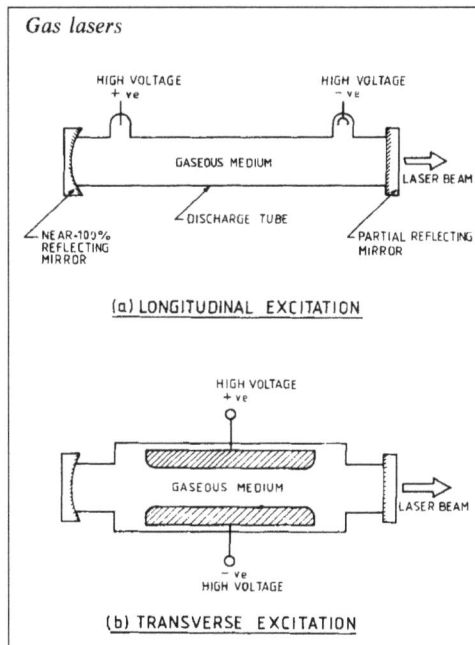

Essential elements of a gas laser.

The electric field which produces the discharge is usually applied along the axis of the laser tube and this configuration is called longitudinal excitation. In some cases, particularly in high-pressure gas lasers, the electric field is applied perpendicular to the laser axis, and this is known as transverse excitation. In this configuration the voltage required for maintaining the discharge is less than that which would be required in longitudinal excitation and a larger gas volume can be uniformly excited. In order to couple the electric power in the gaseous medium suitable discharge electrodes are incorporated in the laser tube. In the electrical discharge the electrons which maintain the discharge current collide with gas atoms or molecules and either directly excite the lasing species to upper laser level or excite other gases which are mixed with the lasing species and they, in turn, transfer their energy to the lasing species through collisions. Thus population inversion is created and laser action follows.

In gas lasers the efficiency, which is defined as optical power output divided by input electrical power, ranges from less than 1% to as high as 30%. Thus only a part of the electrical power is coupled out as useful optical radiation and the rest is dissipated in the medium as heat. This raises the temperature of the medium. For each type of laser there is a certain range of gas temperature beyond which the laser output starts falling. This problem is particularly severe in high-power CW lasers. In low-power, sealed-off lasers heat is removed through the wall of the laser tube by either natural or forced air-cooling, or with water made to flow through a jacket around the tube. In medium- and high-power lasers the gaseous media themselves are continuously made to flow out of the lasing zone and replaced with fresh cooled gas, or are recirculated after cooling through a heat exchanger.

As mentioned earlier some lasers are operated in pulse mode only. They are not suitable for CW operation. The reason could be one or both of the following:

1. The lower laser level where the laser transition terminates is long-lived compared to the upper laser level. During laser action, this level gets populated and when this population is equal to that of the upper laser level the laser action is self-terminated.

2. The lifetime of the upper laser level is very short, of the order of nanoseconds, which makes continuous pumping very difficult. Population inversion in such lasers is established by high-voltage impulses or high-energy electron beam pulses.

Helium-neon Laser

Helium-Neon laser is a type of gas laser in which a mixture of helium and neon gas is used as a gain medium. Helium-Neon laser is also known as He-Ne laser. The helium-neon laser was the first continuous wave (CW) laser ever constructed. It was built in 1961 by Ali Javan, Bennett, and Herriott at Bell Telephone Laboratories.

Helium-neon lasers are the most widely used gas lasers. These lasers have many industrial and scientific uses and are often used in laboratory demonstrations of optics. In He-Ne lasers, the optical pumping method is not used instead an electrical pumping method is used. The excitation of electrons in the He-Ne gas active medium is achieved by passing an electric current through the gas. The helium-neon laser operates at a wavelength of 632.8 nanometers (nm), in the red portion of the visible spectrum.

Helium-neon Laser Construction

The helium-neon laser consists of three essential components:

- Pump source (high voltage power supply).

- Gain medium (laser glass tube or discharge glass tube).

- Resonating cavity.

High Voltage Power Supply or Pump Source

In order to produce the laser beam, it is essential to achieve population inversion. Population inversion is the process of achieving more electrons in the higher energy state as compared to the lower energy state. In general, the lower energy state has more electrons than the higher energy state. However, after achieving population inversion, more electrons will remain in the higher energy state than the lower energy state.

In order to achieve population inversion, we need to supply energy to the gain medium or active medium. Different types of energy sources are used to supply energy to the gain medium.

In ruby lasers and Nd:YAG lasers, the light energy sources such as flashtubes or laser diodes are used as the pump source. However, in helium-neon lasers, light energy is not used as the pump source. In helium-neon lasers, a high voltage DC power supply is used as the pump source. A high voltage DC supplies electric current through the gas mixture of helium and neon.

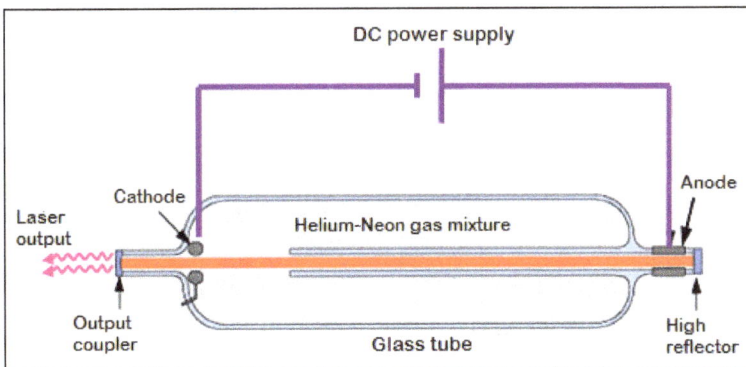

Gain Medium (Discharge Glass Tube or Glass Envelope)

The gain medium of a helium-neon laser is made up of the mixture of helium and neon gas contained in a glass tube at low pressure. The partial pressure of helium is 1 mbar whereas that of neon is 0.1 mbar. The gas mixture is mostly comprised of helium gas. Therefore, in order to achieve population inversion, we need to excite primarily the lower energy state electrons of the helium atoms.

In He-Ne laser, neon atoms are the active centers and have energy levels suitable for laser transitions while helium atoms help in exciting neon atoms. Electrodes (anode and cathode) are provided in the glass tube to send the electric current through the gas mixture. These electrodes are connected to a DC power supply.

Resonating Cavity

The glass tube (containing a mixture of helium and neon gas) is placed between two parallel mirrors. These two mirrors are silvered or optically coated. Each mirror is silvered differently. The left side mirror is partially silvered and is known as output coupler whereas the right side mirror is fully silvered and is known as the high reflector or fully reflecting mirror.

The fully silvered mirror will completely reflect the light whereas the partially silvered mirror will reflect most part of the light but allows some part of the light to produce the laser beam.

Working of Helium-neon Laser

In order to achieve population inversion, we need to supply energy to the gain medium. In helium-neon lasers, we use high voltage DC as the pump source. A high voltage DC produces energetic electrons that travel through the gas mixture. The gas mixture in helium-neon laser is mostly comprised of helium atoms. Therefore, helium atoms observe most of the energy supplied by the high voltage DC.

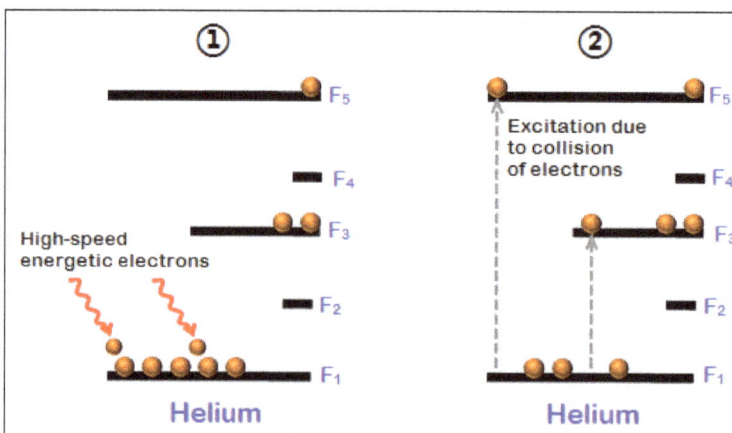

When the power is switched on, a high voltage of about 10 kV is applied across the gas mixture. This power is enough to excite the electrons in the gas mixture. The electrons produced in the process of discharge are accelerated between the electrodes (cathode and anode) through the gas mixture.

In the process of flowing through the gas, the energetic electrons transfer some of their energy to the helium atoms in the gas. As a result, the lower energy state electrons of the helium atoms gain enough energy and jumps into the excited states or metastable states. Let us assume that these metastable states are F_3 and F_5.

The metastable state electrons of the helium atoms cannot return to ground state by spontaneous emission. However, they can return to ground state by transferring their energy to the lower energy state electrons of the neon atoms.

The energy levels of some of the excited states of the neon atoms are identical to the energy levels of metastable states of the helium atoms. Let us assume that these identical energy states are $F_3 = E_3$ and $F_5 = E_5$. E_3 and E_5 are excited states or metastable states of neon atoms.

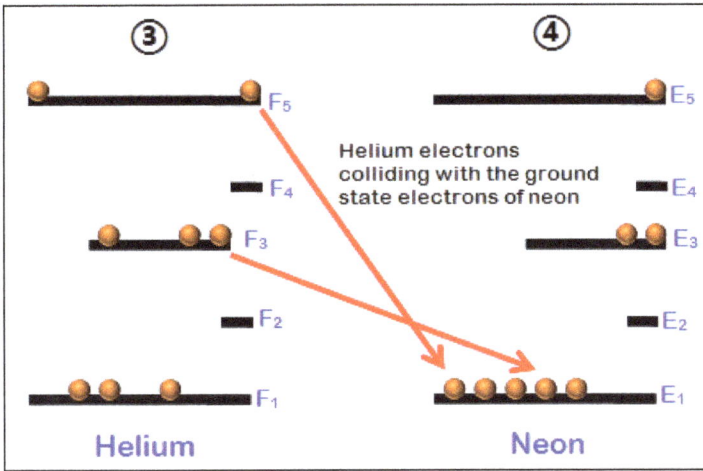

Unlike the solid, a gas can move or flow between the electrodes. Hence, when the excited electrons of the helium atoms collide with the lower energy state electrons of the neon atoms, they transfer their energy to the neon atoms. As a result, the lower energy state electrons of the neon atoms gain enough energy from the helium atoms and jumps into the higher energy states or metastable states (E_3 and E_5) whereas the excited electrons of the helium atoms will fall into the ground state. Thus, helium atoms help neon atoms in achieving population inversion.

Likewise, millions of ground state electrons of neon atoms are excited to the metastable states. The metastable states have the longer lifetime. Therefore, a large number of electrons will remain in the metastable states and hence population inversion is achieved.

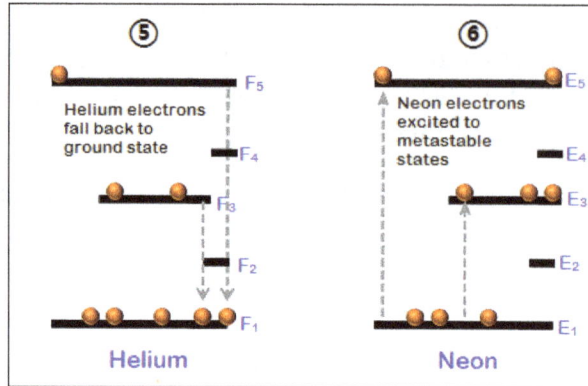

After some period, the metastable states electrons (E_3 and E_5) of the neon atoms will spontaneously fall into the next lower energy states (E_2 and E_4) by releasing photons or red light. This is called spontaneous emission. The neon excited electrons continue on to the ground state through radiative and nonradiative transitions. It is important for the continuous wave (CW) operation.

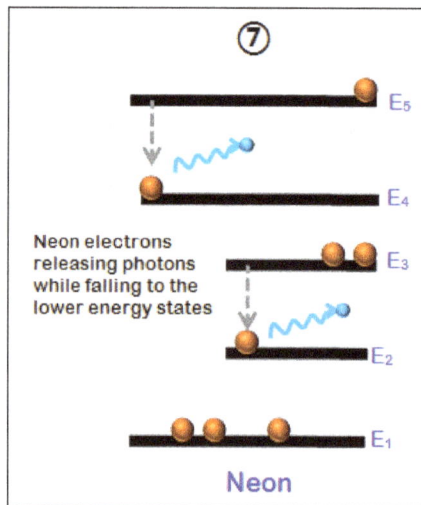

The light or photons emitted from the neon atoms will moves back and forth between two mirrors until it stimulates other excited electrons of the neon atoms and causes them to emit light. Thus, optical gain is achieved. This process of photon emission is called stimulated emission of radiation. The light or photons emitted due to stimulated emission will escape through the partially reflecting mirror or output coupler to produce laser light.

Advantages of Helium-neon Laser

- Helium-neon laser emits laser light in the visible portion of the spectrum;

- High stability;

- Low cost;

- Operates without damage at higher temperatures.

Disadvantages of Helium-neon Laser

- Low efficiency;

- Low gain;

- Helium-neon lasers are limited to low power tasks.

Applications of Helium-neon Laser

- Helium-neon lasers are used in industries.

- Helium-neon lasers are used in scientific instruments.

- Helium-neon lasers are used in the college laboratories.

Carbon Dioxide Laser

The CO_2 laser (*carbon dioxide laser*) is a molecular gas laser based on a gas mixture as the gain medium, which contains carbon dioxide (CO_2), helium (He), nitrogen (N_2), and possibly some hydrogen (H_2), water vapor and/or xenon (Xe). Such a laser is electrically pumped via a gas discharge, which can be operated with DC current, with AC current (e.g. 20–50 kHz) or in the radio frequency (RF) domain. Nitrogen molecules are excited by the discharge into a metastable vibrational level and transfer their excitation energy to the CO_2 molecules when colliding with them. Helium serves to depopulate the lower laser level and to remove the heat. Other constituents such as hydrogen or water vapor can help (particularly in sealed-tube lasers) to reoxidize carbon monoxide (formed in the discharge) to carbon dioxide.

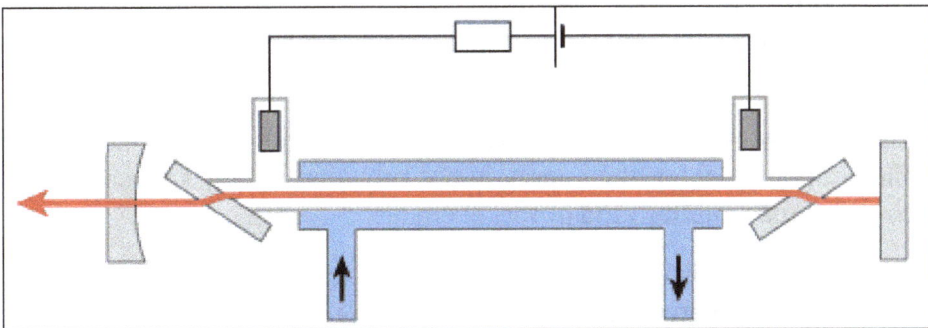

Schematic setup of a sealed-tube carbon dioxide laser.
The gas tube has Brewster windows and is water-cooled.

CO_2 lasers typically emit at a wavelength of 10.6 µm, but there are other lines in the region of 9–11 µm (particularly at 9.6 µm). In most cases, average powers are between some tens of watts and many kilowatts. The power conversion efficiency can be well above 10%, i.e., it is higher than for most gas lasers (due to a particularly favorable excitation pathway), also higher than for lamp-pumped solid-state lasers, but lower than for many diode-pumped lasers.

Due to their high output powers, CO_2 lasers require high-quality infrared optics, often made of materials like zinc selenide (ZnSe) or zinc sulfide (ZnS).

Laser Types

The family of CO_2 lasers is very diverse:

- For laser powers between a few watts and a several hundred watts, it is common to use sealed-tube or no-flow lasers, where the laser bore and gas supply are contained in a sealed tube. Such lasers are compact and rugged, and reach operation lifetimes of several thousands of hours.

- High-power diffusion-cooled slab lasers have the gas in a gap between a pair of planar water-cooled RF electrodes. The excess heat is efficiently transferred to the electrodes by diffusion, if the electrode spacing is made small compared with the electrode width. Several kilowatts of output are possible.

- Fast axial flow lasers and fast transverse flow lasers are also suitable for multi-kilowatt continuous-wave output powers. The excess heat is removed by the fast-flowing gas mixture, which passes an external cooler before being used again in the discharge.

- Transverse excited atmosphere (TEA) lasers have a very high (about atmospheric) gas pressure. As the voltage required for a longitudinal discharge would be too high, transverse excitation is done with a series of electrodes along the tube. TEA lasers are operated in pulsed mode only, as the gas discharge would not be stable at high pressures. They often produce average output powers below 100 W, but can also be made for powers of tens of kilowatts (combined with high pulse repetition rates).

- There are gas dynamic CO_2 lasers for multi-megawatt powers (e.g. for anti-missile weapons), where the energy is not provided by a gas discharge but by a chemical reaction in a kind of rocket engine.

The concepts differ mainly in the technique of heat extraction, but also in the gas pressure and electrode geometry used. In low-power sealed-tube lasers (used e.g. for laser marking), waste heat is transported to the tube walls by diffusion or a slow gas flow. The beam quality can be very high. High-power CO_2 lasers utilize a fast forced gas convection, which may be in the axial direction (i.e., along the beam direction) or in the transverse direction (for the highest powers).

Applications

CO_2 lasers are widely used for material processing, in particular for:

- Cutting plastic materials, wood, die boards, etc., exhibiting high absorption at 10.6 μm, and requiring moderate power levels of 20–200 w.

- Cutting and welding metals such as stainless steel, aluminum or copper, applying multi-kilowatt powers.

- Laser marking of various materials.

Other applications include laser surgery (including ophthalmology) and range finding. CO_2 lasers used for material processing (e.g. welding and cutting of metals, or laser marking) are in competition with solid-state lasers (particularly YAG lasers and fiber lasers) operating in the 1-μm wavelength regime. These shorter wavelengths have the advantages of more efficient absorption in a metallic workpiece, and the potential for beam delivery via fiber cables. (There are no optical fibers for high-power 10-μm laser beams). The potentially smaller beam parameter product of 1-μm lasers can also be advantageous. However, the latter potential normally cannot be realized with high-power lamp-pumped lasers, and diode-pumped lasers tend to be more expensive. For these reasons, CO_2 lasers are still widely used in the cutting and welding business, particularly for parts with a thickness greater than a few millimeters, and their sales make more than 10% of all global laser sales (as of 2013). This may to some extent change in the future due to the development of high-power thin-disk lasers and advanced fiber cables in combination with techniques which exploit the high beam quality of such lasers.

Due to their high powers and high drive voltages, CO_2 lasers raise serious issues of laser safety. However, their long operation wavelength makes them relatively eye-safe at low intensities.

Nitrogen Laser

A nitrogen laser is a gas laser operating in the ultraviolet (UV) range by using molecular nitrogen as its gain medium. Nitrogen lasers were first developed in 1963, and began to be used commercially in 1972.

Nitrogen lasers operate based on a fast electrical discharge through nitrogen gas. The nitrogen gas can be supplied through a gas cylinder, or from liquid nitrogen. The laser light emitted is in the UV range, with a short pulse width and high intensity. The nitrogen laser uses electricity to excite the nitrogen. When an electric spark crosses a spark gap in the laser, the electrons hit the nitrogen atoms in air thereby exciting them into a metastable state. When a photon with a wavelength of 337 nm passes the excited nitrogen atoms, stimulated emission occurs and a laser state is generated.

Nitrogen lasers find applications in research in medicine, chemistry and physics.

Laser Properties	
Laser type	Gas
Pump source	Electrical discharge
Operating Wavelength	337.1 nm

Nitrogen lasers can be used for a wide range of applications in the UV-visible region. They can be easily coupled to a microscope for carrying out experiments in life science laboratories. They are also efficient sources for laser-induced fluorescence and photochemistry and general spectroscopy.

Other major applications of nitrogen lasers include:

- Measurement of air pollution.

- Treatment of nonhealing wounds, pulmonary tuberculosis, etc.

- Transverse optical pumping of dye lasers.

Construction and Operation

Like all lasers, nitrogen lasers typically consist of three basic parts: An energy source (or pump), a laser medium (also known as a gain medium), and an optical resonator.

- As in most gas lasers, the energy source is an electrical discharge provided by a power supply.

- The gain medium is some concentration of N_2 molecules. The laser medium is typically either pure nitrogen, nitrogen-helium mixture, or simply air.

- Unlike many lasers, nitrogen lasers can operate without an optical resonator (the series of mirrors and windows used to amplify and direct emitted radiation). This is due to the fact that stimulation of nitrogen atoms results in amplified spontaneous emission (SAE) — also known as superluminescent light — by achieving population inversion within the gain medium. (Population inversion refers to a state where more atoms exist in an excited state than in a low energy state.) Nitrogen lasers may still include a single reflective mirror at the back of the laser to ensure correct output.

Nitrogen lasers use simpler and cheaper building materials when compared to more powerful gas lasers such as excimer and CO_2 lasers. Because the gas medium is relatively benign, the gas is often contained within a simple plastic (acrylic glass) chamber. The image below shows a typical home-built (non-commercial) nitrogen laser reflecting the simplicity of the parts involved. In this design, the spark gap at right provides the electrical discharge which is transmitted through the plates and into the gas chamber. The chamber itself is made of Plexiglas, while the sole optical components are a microscope slide and covers positioned at the output point.

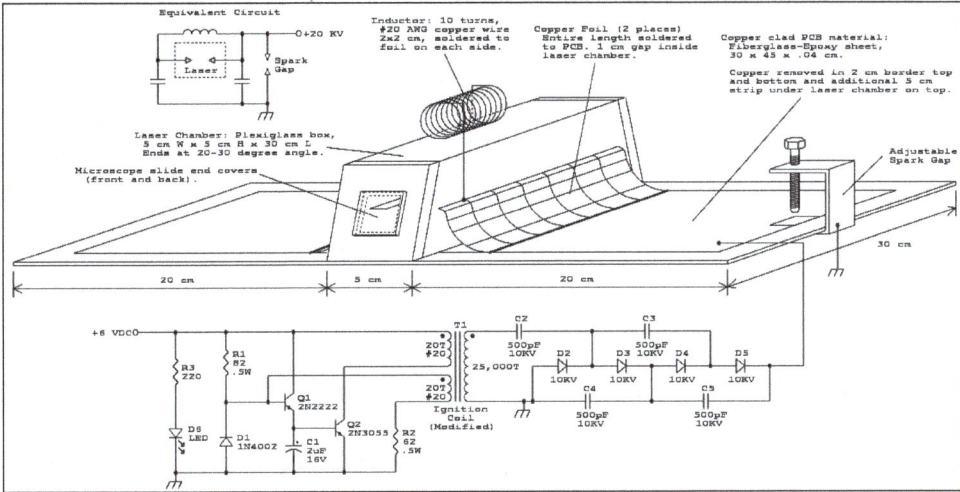

Typical Home-Built Nitrogen Laser Assembly.

Nitrogen lasers are exclusively ultraviolet (UV) devices which predominantly emit at 337.1 nm. Their operation consists solely of extremely short, powerful pulses.

Levels and Population Inversion

Nitrogen lasers achieve superluminescent emission by using population inversion. In order to accomplish this, they use three different energy levels (and are therefore classified as three-level lasers). It is important to note that only two of the energy levels are specifically important, as the third level represents the unexcited ground state.

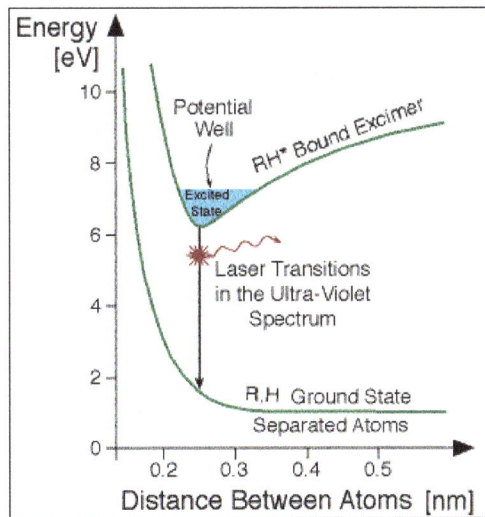

Simplified energy level diagram of Nitrogen Laser.

A nitrogen laser's upper energy level is directly pumped by a high voltage electrical discharge. This pumping is powerful enough to cause population inversion and stimulates the nitrogen atoms to immediately emit at 337.1 nm; the atoms remain at this level for approximately 20 ns. As the energy level drops the atoms fall into the second level and remain there for much longer (approximately 10 microseconds) than the first level. The effect of this time difference is that the atoms rapidly cease emission, resulting in a fast pulse of radiation. When considering this effect, it is clear that the pulse length is directly determined by the lifetime of the upper level, which in turn is dependent upon the gas medium's pressure: the higher the pressure, the shorter the lifetime of the level (and pulse).

The graph below shows the three energy levels (as well as a fourth "meta-stable" level) of a typical nitrogen laser. The purple line represents the lasing action, which is incited by the pumping of the upper level and terminated at the second level.

Gas Pressure and TEA Lasers

Nitrogen lasers are generally constructed using one of two designs. The first is a "traditional" low-pressure design involving a vacuum pump; this one results in pulses 5-10 ns in length and is most suitable for pumping other lasers.

The second design is a higher-pressure one known as a TEA (transverse electrical discharge at atmospheric pressure) nitrogen laser. (It is important to note that, while TEA lasers are frequently nitrogen types, TEA designs for excimer and carbon dioxide lasers are common as well.) Nitrogen TEA lasers use plain "open" air at atmospheric pressure as their lasing medium, as air is 78% nitrogen. While TEA lasers require no vacuum pump to operate, they require a much faster electrical discharge to achieve effective pulses when compared to lower pressure designs. The home-built laser diagram above represents a TEA laser.

Applications

Compared to other gas lasers, nitrogen lasers are used in a relatively narrow range of applications. Because of their simple construction and inexpensive components, they are highly valued by beginning laser hobbyists. Practical nitrogen laser uses include:

- Nondestructive testing (NDT);
- Measurement of rapid processes such as time of flight (TOF), due to fast pulsing;
- Pumping of dye lasers;
- UV spectroscopy;
- Fluorescence measurement.

Gas Dynamic Laser

A gas dynamic laser (GDL) is a laser based on differences in relaxation velocities of molecular vibrational states. The lasing medium gas has such properties that an energetically lower vibrational state relaxes faster than a higher vibrational state, and so a population inversion is achieved in a particular time. It was invented by Edward Gerry and Arthur Kantrowitz at Avco Everett Research Laboratory in 1966.

Pure gas dynamic lasers usually use a combustion chamber, supersonic expansion nozzle, and.CO, in a mixture with nitrogen or helium, as the laser medium. Gas dynamic lasers can be pumped by combustion or adiabatic expansion of gas. Any hot and compressed gas with appropriate vibrational structure could be utilized.

The explosively pumped gas dynamic laser is a version of GDL pumped by expansion of explosion products. Hexanitrobenzene and/or tetranitromethane with metal powder is the preferred explosive. This device could have very high pulsed peak power output suitable for laser weapons.

Functions of GDL

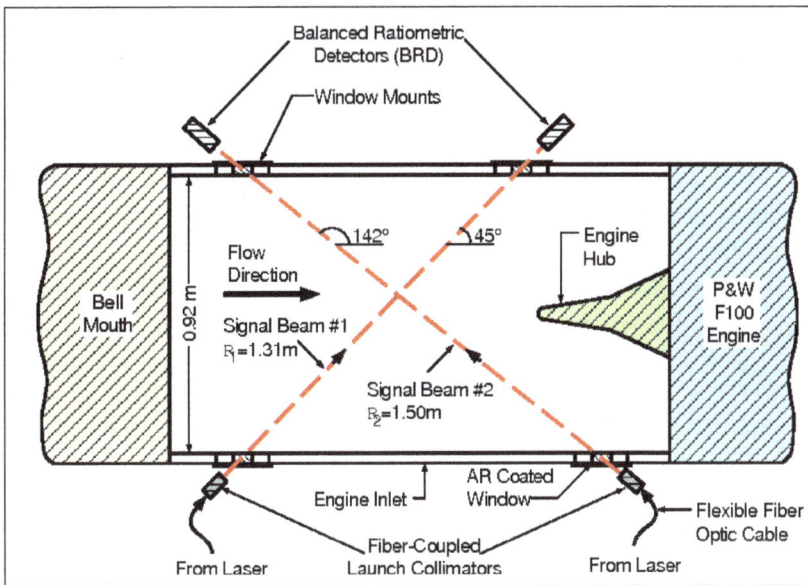

Gas dynamic laser components and function.

- Hot compressed gas is generated.

- Gas expands through subsonic or supersonic expansion nozzle, the temperature of the gas becomes lower and according to Maxwell–Boltzmann distribution the gas isn't in thermodynamic equilibrium until the vibrational states relax.

- The gas flows through the tube of a particular length for a particular time. In this time lower vibrational state does relax but higher vibrational state doesn't. Thus population inversion is achieved.

- Gas flows through mirror area where stimulated emission takes place.

- Gas returns to equilibrium and becomes warm. It must be removed from the laser cavity or it will interfere with the thermodynamics and vibrational state relaxation of the freshly expanded gas.

Application

- Almost any chemical laser uses gas-dynamic processes to increase its efficiency.

- High energy efficiency (as high as 30%) and very high power output make GDL suitable for some (especially military) applications.

Excimer Laser

An excimer laser is a powerful kind of laser which is nearly always operated in the ultraviolet (UV) spectral region ($\rightarrow$ ultraviolet lasers) and generates nanosecond pulses. The excimer gain medium is a gas mixture, typically containing a noble gas (rare gas) (e.g. argon, krypton, or xenon) and a halogen (e.g. fluorine or chlorine, e.g. as HCl), apart from helium and/or neon as buffer gas. An excimer gain medium is pumped with short (nanosecond) current pulses in a high-voltage electric discharge (or sometimes with an electron beam), which create so-called excimers (excited dimers) – molecules which represent a bound state of their constituents only in the excited electronic state, but not in the electronic ground state. (More precisely, a dimer is a molecule consisting of two equal atoms, but the term excimer is normally understood to include asymmetric molecules such as XeCl as well. The term rare gas halide lasers would actually be more appropriate, and the term exciplex laser is sometimes used.) After stimulated or spontaneous emission, the excimer rapidly dissociates, so that reabsorption of the generated radiation is avoided. This makes it possible to achieve a fairly high gain even for a moderate concentration of excimers.

As excimer lasers use molecules as the gain medium, they can also be called molecular lasers.

Table: Different types of excimer lasers typically emit at wavelengths between 157 and 351 nm:

Excimer	Wavelength
F_2 (fluorine)	157 nm
ArF (argon fluoride)	193 nm
KrF (krypton fluoride)	248 nm
XeBr (xenon bromide)	282 nm
XeCl (xenon chloride)	308 nm
XeF (xenon fluoride)	351 nm

For various of those wavelengths, specialized excimer optics (ultraviolet optics) have been developed, which need to have a high optical quality and in particular a very high resistance to the intense ultraviolet light.

Typical excimer lasers emit pulses with a repetition rate up to a few kilohertz and average output powers between a few watts and hundreds of watts, which makes them the most powerful laser sources in the ultraviolet region, particularly for wavelengths below 300 nm. The wall-plug efficiency varies between 0.2% and 2%.

Device Lifetime

Early excimer lasers had limited lifetimes due to a variety of problems, arising e.g. from the corrosive nature of the gases used and from contamination of the gas with chemical byproducts and dust created by the electric discharge. Other challenges are the ablation of material from the electrodes and the high peak power of the required current pulses, which often allowed the thyratron switches to last only for a couple of weeks or months. However, a lot of engineering, involving e.g. the use of corrosion-resistant materials, advanced gas recirculating and purification systems, and solid-state high-voltage switches, has mitigated challenges of the excimer laser concept to a significant extent. The lifetime of modern excimer lasers is now limited by that of the ultraviolet optics, which have to withstand high fluxes of short-wavelength radiation, to something of the order of a few billion pulses.

Applications

The short wavelengths in the ultraviolet spectral region make possible a number of applications:

- The generation of very fine patterns with photolithographic methods (microlithography), for example in semiconductor chip production.

- Material processing with laser ablation, exploiting the very short absorption lengths of the order of a few micrometers in many materials, so that a moderate pulse fluence of a few joules per square centimeter is sufficient for ablation.

- Pulsed laser deposition.

- Laser marking and microstructuring of glasses and plastics.

- Fabrication of fiber bragg gratings.

- Ophthalmology (eye surgery), particularly for vision correction by corneal reshaping with arf lasers at 193 nm; common methods are laser in-situ keratomileusis (lasik) and photorefractive keratectomy (prk).

- Psoriasis treatment with xecl lasers at 308 nm.

- Pumping other lasers, e.g. Certain dye lasers.

Photolithography in semiconductor device manufacturing is an application of major importance. Here, photoresists on processed semiconductor wafers are irradiated with high-power ultraviolet light through structured photomasks. High-power UV light, as can be generated with excimer lasers, is essential for obtaining short processing times and correspondingly high throughput, while the short wavelengths allow one to fabricate very fine structures (with optimized techniques even far below the optical wavelength). However, the latest developments in lithography require even shorter wavelengths in the extreme ultraviolet (EUV), e.g. at 13.5 nm, which can no longer be produced with excimer lasers. Certain laser-generated plasma sources are developed as the successors for excimer lasers in that area. Still, it is to be expected that excimer lasers will be used for fabricating many semiconductor chips for a long time to come, as only the most advanced computer chips require still finer structures than possible with such techniques.

Chemical Laser

Chemical lasers find their origin in the study of the radiation emitted from chemical reactions. In many cases such radiation is of thermal origin. Thus, much of the radiation from combustion processes is due to the black or grey body radiation of solid particles (soot) made up mostly of carbon. The infrared radiation consists of CO_2 and H_2O emissions which reflect the temperature of the combustion zone. Ultraviolet radiation is weak because most chemical reactions do not produce high enough temperatures to give rise to blackbody ultraviolet radiation. In these cases the energy released by the chemical reactions is observed as thermal energy, and the same radiation could be observed by indirectly heating the reaction products to the appropriate temperature. This is not always the case; perhaps the best known exception is the blue-green emission observed from the oxidation of phosphorus in air, which gave rise to the term "phosphorescence," one definition is: "an enduring luminescence without sensible heat."

Indeed, the phosphorus flame, in spite of its visible radiation, is without heat: a piece of paper held in the flame will not ignite. A more pertinent case is the radiation observed from low pressure flames. Polanyi and his coworkers have studied the infrared emission from the $H + Cl_2$ flame at low pressure and found vibrational populations characteristic of 7350 K, simultaneously with rotational populations indicating a 500 K temperature. An atmospheric pressure flame of H_2-Cl_2, on the other hand, has a temperature of about 2500 K with infrared emission characteristic of this temperature. Clearly, therefore, the HCl produced in the low pressure flame has a vibrational population distribution which is not representative of the gas temperature and many studies have demonstrated conclusively that the product is formed in an excited vibrational state. At low pressures the vibrationally excited HCl has a chance to radiate before the vibrational energy is changed to translational energy and averaged over the system. Rotational relaxation is much faster than vibrational relaxation, i.e., takes fewer collisions; thus, the rotational levels tend to be in equilibrium with the translational gas temperature even at pressures of a few torr. In a one atmosphere flame there are sufficient collisions to relax the vibrationally excited HCl, the mixture reaches thermal equilibrium and the emission is that of HCl at the reaction temperature. Polanyi, and independently Penner, suggested in 1961 that the non-equilibrium distribution resulting from such reactions could be used to "pump" an infrared chemical laser.

A laser, chemical or otherwise, requires a mechanism to populate an excited state at a sufficiently fast rate such that at some time point there are more molecules in an upper (higher energy) state than in a lower. Under these conditions the number of photons produced by stimulated emission can exceed those absorbed, and optical amplification or gain will result. Under equilibrium conditions the ratio of the number of molecules in an upper and lower state are given by the Boltzmann distribution:

$$\frac{N_u}{N_1} = \frac{g_u}{g_1} e - \Delta E / kT$$

where T is the absolute temperature, ΔE is the energy difference between the states, and the g's are the dengenerates of the states, If we ignore the physical meaning of T and examine the behavior of equation above as a function of T, it is clear that for T > 0, at least for a simple two-level system, Nu < N1, i.e., an inversion is not possible for finite, positive T's. Only when T is negative can we have an inversion. This has given rise to the term negative temperature. Similarly, application of equation:

$$\frac{N_u}{N_1} = \frac{g_u}{g_1} e - \Delta E / kT$$

to any two vibrational or rotational levels out of an assembly of levels of a molecule gives rise to such terms as vibrational and rotational "temperatures." These terms are rather unfortunate in that the conventional use of the word temperature implies a condition of thermodynamic equilibrium, whereas in the present case just the opposite is true. When an inversion is achieved by a chemical reaction, the laser is referred to as a chemical laser. Historically, photo dissociation lasers and sometimes gas dynamic lasers have been considered chemical lasers.

The chemical production of excited state species must be fast enough to overcome the losses of excited species, which in the case of infrared molecular lasers, are due mainly to collisons. An examination of these loss processes shows that vibrational-to-translational (V-T) and vibrational-to-rotational (V-R) energy conversion are generally much slower than either vibrational-vibrational (V-V), rotational-rotational (R-R) and rotational-to-translational (R-T) energy transfers. This suggests the possibility of a pseudoequilibrium where the vibrational levels are in equilibrium among themselves due to rapid V-V exchange, but not with the rotational or translational levels because of the much slower V-R and V-T processes. This then leads to the concept of partial inversions characterized by vibrational and rotational "temperatures." For a diatomic molecule, in the harmonic oscillator-rigid rotator approximation the energy levels are given by:

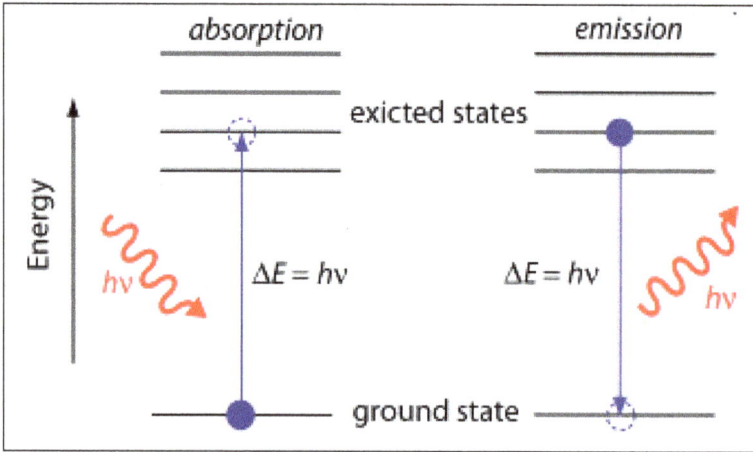

Absorption and emission processes.

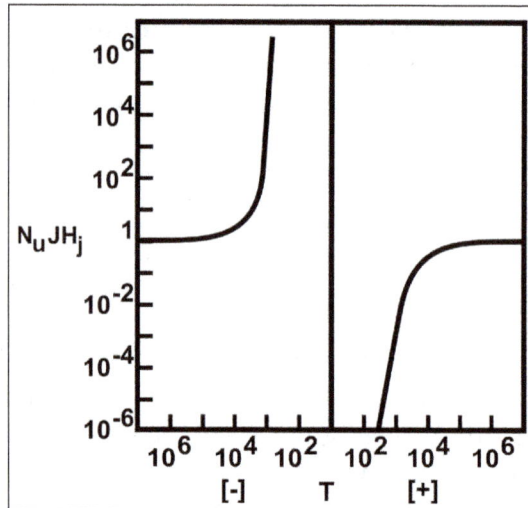

Values of $N_u/N_1 = \exp(-\Delta E/kT)$ for $\Delta E = 4000$ cm^{-1}.

$$E(V,J) = \omega(V + 1/2) BJ(J+1)$$

where ω is the rotationless vibrational frequency, B is the rotational constant, and V and J are vibrational and rotational quantum numbers, respectively. Transitions with ΔV and $\Delta J = \pm 1$ give rise to the familiar R and P branches in the spectra of heterogeneous diatomic molecules. The condition for gain is given by:

$$\frac{N_u}{g_u} > \frac{N_1}{g_1}$$

The number of molecules in a vibrational state is given by:

$$N_V = \frac{N}{Q_V} \exp\left[\frac{-\omega(V+1/2)hc}{kT_V}\right]$$

In any given vibrational level the rotational level populations are given by:

$$N_{V,J} = \frac{N_V}{Q_R} \exp\left[\frac{-J(J+1)Bhc}{kT_R}\right]$$

If the whole system is in equilibrium $T_V = T_R$, under conditions of partial equilibrium T_V and T_R can be different. In general, T_R will be equal to the translational temperature. Substitution of equation:

$$N_V = \frac{N}{Q_V} \exp\left[\frac{-\omega(V+1/2)hc}{kT_V}\right]$$

and

$$N_{V,J} = \frac{N_V}{Q_R} \exp\left[\frac{-J(J+1)Bhc}{kT_R}\right]$$

in

$$\frac{N_u}{g_u} > \frac{N_1}{g_1}$$

Gives:

$$\exp\left[\frac{-\omega hc(V'+1/2)}{kT_V}\right] \exp\left[\frac{-J'(J'+1)Bhc}{kT_R}\right]$$

$$> \exp\left[-whc(V+1/2)\right] \exp\left[\frac{-J(J+1)Bhc}{kT_R}\right]$$

where the primes indicate the upper level. Taking logarithms and letting V' = V + 1 (i.e., V is the lower level, V + 1 the upper):

$$\frac{\omega}{T_V} + \frac{BJ'(J'+1)}{T_R} > \frac{BJ(J+1)}{T_R}$$

or

$$\frac{T_R}{T_V} > \left\{ J(J+1) - J'(J'+1) \right\} \frac{B}{\omega}$$

For P branch transitions: $\Delta J = +1$, i.e., $J' = J + 1$, and

$$\frac{T_R}{T_V} < 2J \frac{B}{\omega}$$

For R branch transitions: $\Delta J = -1$ i.e., $J' = J - 1$, and

$$\frac{T_R}{T_V} < -2(J+1)\frac{B}{\omega}$$

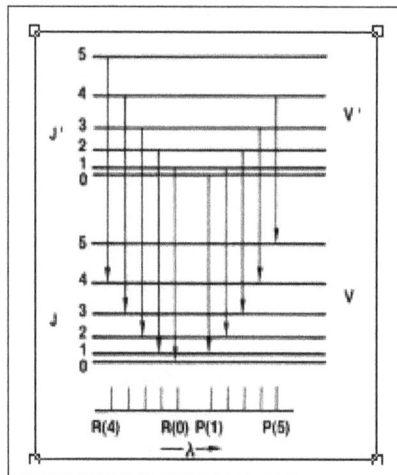

Typical vibration-rotation energy levels and spectrum
for a heterogeneous diatomic molecule.

Since J, B, and ω are all positive, gain in the P-branch can occur with both TV and Tr positive. All that is required is that $T_V > T_R$ by an amount determined by the vibrational and rotational constants and the rotational quantum number. For the R-branch, on the other hand, a negative Ty or total inversion among the vibrational levels is required. Thus, we see that lasing can occur when there is a non-equilibrium between vibration and rotation, a requirement that is much easier to achieve than a total inversion. In accordance with the above concepts, most molecular lasers operate on the P-branch lines only. Even those with a total vibrational inversion have output spectra with rotational distributions characteristic of the gas temperature, an indication that the concept of separate vibrational and rotational manifolds is generally valid. Figure shows the calculated gain in HF at one torr for the indicated ratios of the V = 0 and V = 1 levels. For $N_1 / N_0 < 1$, only the P-branches show gain: the R-branches show absorption. Only when $N_1 / N_0 > 1$ ($T_V < 0$) is there gain on the R-branch transitions.

Earlier we defined a chemical laser as one in which the inversion is produced by a chemical reaction. Ideally, one would like to reacts two stable chemicals and obtain lasing without the need for auxiliary electrical input. In reality this has been achieved only recently in combustion driven HF, HC1, HBr, and CO lasers. Most chemical lasers require atomic species as at least one of the reagents, which on a small scale can be produced easiest by a pulsed or cw discharge. In other instances photo dissociation is used either to prepare the atoms or to cause photo dissociation into excited species.

The first molecular chemical laser was reported by Pimentel in 1965. It was obtained from the flashlamp initiated reaction between H_2 and Cl_2. Shortly after this initial work, a large number of pulsed HC1 and particularly HF lasers were reported. The corresponding cw lasers were not reported until 1969 and 1970.

Hydrogen Fluoride Laser

The hydrogen fluoride (HF) laser is an infrared chemical laser that combines hydrogen produced in a combustion chamber with fluorine gas produced by thermal decomposition of compounds to create excited HF molecules.

It has the potential to provide up to several megawatts of continuous output. The laser outputs light at a wavelength of 2.7-2.9 μm. This wavelength can be absorbed by the air, thereby attenuating the beam and minimizing its range, and hence it has to be used in a vacuum environment.

Laser Properties	
Laser type	Chemical
Pump source	Chemical reaction in the combustion of nitrogen trifluoride and ethylene
Operating wavelength	2.7 to 2.9 μm

Applications

HF lasers have been primarily used in military and space applications. They are widely used for space-based missile defense systems as the propagation of infrared light is better through upper regions.

Deuterium Laser

The deuterium fluoride (DF) laser is a chemical laser formed by a mixture of fluorine and deuterium gas under controlled conditions. The wavelength of light produced by DF laser is longer than that of other conventional HF lasers, and hence it facilitates more effective laser transmission.

During its operation, ethylene is burned in the presence of nitrogen trifluoride in a combustion chamber, which produces free excited fluorine radicals. The mixture of deuterium and helium gas is then injected into an exhaust stream via a nozzle. The deuterium molecules react with the fluorine radicals to produce excited deuterium fluoride. The excited molecules further undergo stimulated emission in the optical resonating laser region.

Laser Properties	
Laser type	Chemical
Pump source	Chemical reaction
Operating wavelength	~3800 nm

Applications

Since their invention in 1970, DF lasers have been widely used in military applications for developing air and missile defense weapon systems of high power because of their ability to rapidly discard waste heat by convective flow of exhaust gases and store high levels of energy.

The pulsed energy projectile and the tactical high energy lasers and the Mid-infrared advanced chemical laser used in army are of deuterium fluoride type.

Chemical Oxygen Iodine Laser

Chemical oxygen iodine laser (COIL) is an infrared chemical laser that has been developed at the Air Force Research Laboratory in 1977, primarily for military applications. It outputs a beam of infrared at a wavelength of 1.315 μm. In continuous mode, the output power of the laser can be scaled up to megawatts.

The key properties of COIL include:

- Possibility of high power emission based on the chemical reaction.

- Emission of high quality light with its low pressure gas laser medium.

- Absorption of large amount of energy on the material due to its short wavelength.

- Very low transmission loss in optical fibers.

The laser consists of an aqueous mixture of potassium hydroxide and hydrogen peroxide, molecular iodine and gaseous chlorine. When the aqueous peroxide solution is subjected to chemical reaction, it releases heat in addition to the production of potassium chloride molecules and excited oxygen known as singlet delta oxygen. Following this, excited oxygen gains a spontaneous lifetime of about 45 min, which in turn allows the singlet delta oxygen to transfer its energy to the iodine molecules that are incorporated to the gas stream. Further, the excited iodine undergoes stimulated emission in the optical resonating region of laser at 1.315 μm.

This laser is operated at very low gas pressures and rapid gas flow rate in order to eliminate heat from the lasing medium easily when compared to other high-power solid-state lasers. A halogen scrubber is employed to remove the traces of iodine and chlorine from the exhaust gas.

Although COIL was initially developed for military purposes, its properties make it useful for industrial processing applications also.

Laser Properties	
Laser type	Chemical
Pump source	Chemical reaction by combustion of singlet delta oxygen and iodine
Operating wavelength	1.315 µm

Applications

COIL technology has gained importance over the past few years owing to its nearly diffraction-limited optical quality, high scalability and short, fiber-deliverable wavelength. These distinct characteristics have made COIL a suitable technology for nuclear warhead dismantlement and nuclear reactor decommissioning. It is used as a weapon laser for the advanced tactile laser programs and military airborne laser. In addition, COIL can also be used for commercial or industrial applications which demands non-invasive, precise and rapid drilling or cutting.

Other applications of COIL include:

- Paint stripping;

- Rock crushing;

- Dismantling of nuclear facilities;

- Cleanup and survivor rescue during disaster;

- High power fiber optic transmission.

The Copper Vapour Laser

The copper vapour (neutral metal) laser is the most useful in the class of pulsed metal vapour lasers. The primary wavelengths for this laser are 510 and 578nm, and over 100W can be generated in the green and yellow part of the visible spectrum. The copper vapour laser is unusual with respect to its high power and high efficiency in this region and in that its normal operation is at pulse repetition rates of several tens of kilohertz. (The internal physics of this laser prevents CW (Continuous Wave) operation) In neutral metal vapour lasers, a fast electric discharge directly excites metal atoms, the high repetition rates permitting high average power output. The copper vapour comes from pieces of copper placed in the discharge tube, which is heated to about 1500 °C to produce vapour at about 0.1mbar pressure. Several mbars of neon are added as a buffer gas to prevent window contamination and loss of copper. Overall wall plug efficiency is about 1% for these lasers, the highest for visible gas lasers. Figure shows the schematic construction of a copper vapour laser.

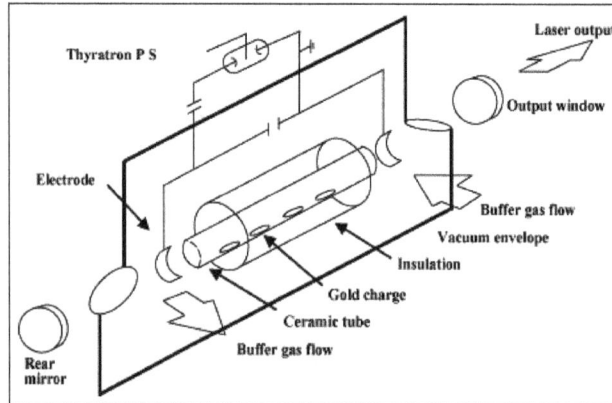

Copper Vapour laser schematic.

The copper vapour laser tube is usually sealed with flat glass windows the rear mirror is a total reflector, with 90% transmission chosen for the output window, which does not need to be specially coated. Beams from copper vapour lasers can be from about 10mm diameter to 50mm diameter, beam divergence for a stable (output window) type resonator is 3-5mradians. In the visible part of the spectrum, beam focusing is performed using glass optics.

Helium Cadium Laser

The Helium-Cadmium (HeCd) laser is is one of a class of gas lasers using helium in conjunction with a metal which vaporizes at a relatively low temperature. Other examples are the Helium-Mercury (HeHg) and Helium-Selenium (HeSe) lasers, which are among those that can be built by a determined amateur.

The typical HeCd laser can produce a high quality beam at 442 nm (violet-blue) and/or 325 nm (UV) depending on the optics. Typical power output is in the 10s to 100s of milliwatts range. In terms of popularity, the HeCd laser probably ranks behind HeNe, Ar/Kr ion, and CO_2 gas lasers. Although its wavelengths may be highly desirable for some forms of spectroscopy and non-destructive testing, they are pretty useless for laser shows and other common hobbyist applications. For that reason, as well as the higher complexity (and cost), one doesn't see these lasers nearly as often as the more common types.

HeCd laser tubes ARE more complex than those used for HeNe, Ar/Kr ion, and CO_2 lasers. In addition to often using a heated filament/cathode, they also include a reservoir for the cadmium metal and a heater to control its vapor pressure, an overall heater and thermal insulation to control tube and helium temperature/pressure, and various sensors inside the envelope to monitor these parameters for use by several feedback loops in the power supply. Even if the mirrors are internal, they are often adjustable to some extent.

HeCd Laser Tube

Although the HeCd is still a gas laser, its construction is quite complex compared to, say, a common HeNe laser tube:

- In addition to the gas fill of helium, there is a cadmium reservoir and heater to maintain a specific (close loop controlled) Cd vapor pressure in the tube. One of the factors determining HeCd tube lifetime is how long enough Cd remains in the reservoir.

- Overall helium pressure is regulated via closed loop feedback utilizing multiple sensors inside the tube envelope providing information to control current to an overall tube heater.

- Since HeCd tubes operate at relatively high current - at least compared to Helium-Neon (HeNe) lasers, a heated filament/cathode is often used rather than the cylindrical design of a HeNe laser.

- HeCd laser tubes often have an internal mirror for just the High Reflector (HR) or for the Output Coupler (OC) mirrors as well. The former arrangement allows line wavelength selection by changing an external OC mirror or via a line selecting prism. Even with internal mirrors, some adjustment is normally provided via compliant mirror mounts with accessible screws.

WARNING: Since the temperature of the tube affects helium and cadmium pressure, both critical for proper operation, the entire tube is usually covered with thermal insulation. In older lasers, this was asbestos. Thus, HAZMAT handling procedures apply any time maintenance or modifications are being done inside the laser head.

Larger HeCd Laser Tube

This is one big Helium Cadmium Laser tube, measures 32.5" long. Made for the typical 3 rod resonator frame. The typical bright gold/yellow color of the discharge with no pink from air leakage visible.

This tube goes in the model 374 laser head which is a larger version of the 456 head. The head's model number indicates the lasing wavelength: A 374 is 325 nm (near-UV) and a 474 or 456 is the 442 nm (violet). Omnichrome is now a part of Melles Griot.

Here are some of the specifications for the tube:

- Output wavelength: 325 nm (near UV) and/or 442 nm (violet).

- Operating voltage: 2.4 to 2.6 kV DC.

- Operating current: 90 to 120 mA.

- Starting voltage: 10 kV DC.

- Helium heater voltage: 6.5 VAC.

- Cadmium heater voltage: 2 to 3 VAC.

- Re-melt heater voltage: 6 to 7 VAC (optional).

- Overall Length: 32.5 inches.

Ion Laser

Ion lasers consist primarily of lasers operating using ionoised species of the noble gases argon, krypton or xenon. Argon with strong emissions in the visible blue green and weaker lines in the ultra violet and infrared is the most important type commercially. The attraction of the argon-ion laser is the ability to produce CW (continuous wave) output from mW to tens of W in the visible part of the spectrum. The technology however, is fairly complex as can be seen from the schematic representation in figure.

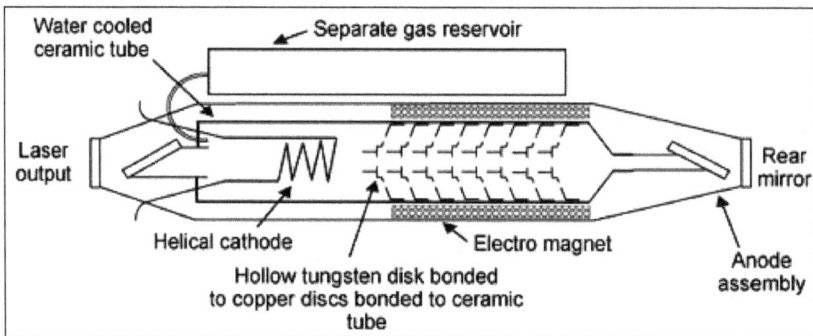

Argon ion laser schematic.

Argon-ion lasers operate in high temperature plasma tubes with bores about 1.2mm diameter, and lengths up to 1.5m. Excitation is by a high current discharge that passes along the length of the tube, concentrated in the small bore. High current density in the centre of the tube ionises the gas and provides the energy to excite the ion to the lasing energy levels. The high current density leads to sputting of the bore materials by the plasma, which is generally detrimental. Extra gas is needed to replenish gas depleted during operation and the low efficiency (~0.1% wall plug) requires methods of removing waste heat. Solutions to these problems use high temperature ceramics, tungsten, separate gas flow paths and active cooling. In addition, magnetic fields are sometimes used to help conform the discharge current to the centre of the bore.

Fused silicon or crystal quartz are usually used for the optics on these lasers, quartz being used for high power visible and UV applications. Typical values for beam diameters for argon ion lasers are from 0.6 to 2mm with beam divergences of 0.4 to 1.2 mradians.

Argon Ion Laser

Argon ion lasers are powerful gas lasers, which typically generate multiple watts of optical power in a green or blue output beam with high beam quality.

The core component of an argon ion laser is an argon-filled tube, made e.g. of beryllium oxide ceramics, in which an intense electrical discharge between two hollow electrodes generates a plasma with a high density of argon (Ar^+) ions. A solenoid around the tube can be used for generating a magnetic field, which increases the output power by better confining the plasma.

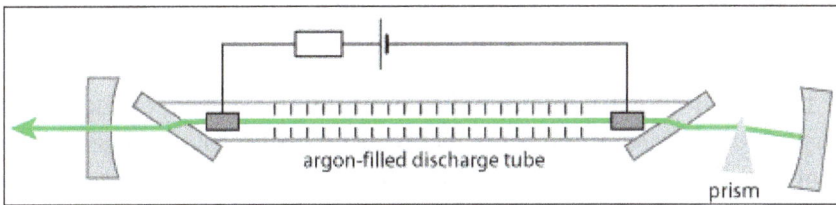

Setup of a 20-W argon ion laser. The gas discharge with high current density occurs between the hollow anode and cathode. The intracavity prism can be rotated to select the operation wavelength.

A typical device, containing a tube with a length of the order of 1 m, can generate 10 W or 20 W of output power in the green spectral region at 514.5 nm, using several tens of kilowatts of electric power. (The voltage drop across the tube may be 100 V or a few hundred volts, whereas the current can be several tens of amperes.) The dissipated heat must be removed with a water flow around the tube; a closed-circle cooling system often contains a chiller, which further adds to the power consumption. The total wall-plug efficiency is thus very low, usually below 0.1%. There are smaller air-cooled argon ion lasers, generating some tens of milliwatts of output power from several hundred watts of electric power.

The laser can be switched to other wavelengths such as 457.9 nm (blue), 488.0 nm (blue–green), or 351 nm (ultraviolet) by rotating the intracavity prism (on the right-hand side). The highest output power is achieved on the standard 514.5-nm line. Without an intracavity prism, argon ion lasers have a tendency for multi-line operation with simultaneous output at various wavelengths.

There are similar noble gas ion lasers based on krypton instead of argon. Krypton ion lasers typically emit at 647.1 nm, 413.1 nm, or 530.9 nm, but various other lines in the visible, ultraviolet and infrared spectral region are accessible.

Applications

Multi-watt argon ion lasers can be used e.g. for pumping titanium–sapphire lasers and dye lasers, or for laser light shows. They are rivaled by frequency-doubled diode-pumped solid-state lasers. The latter are far more power efficient and have longer lifetimes, but are more expensive. Argon tubes have a limited lifetime of the order of a few thousand hours. An argon laser may thus be preferable if it is used only during a limited number of hours, whereas a diode-pumped solid-state laser is the better solution for reliable and efficient long-term operation.

Laser safety issues arise both from the high output power of typical ion lasers and from the high voltage applied to the tube.

Krypton Ion Laser

Krypton laser belongs to the gas lasers family, which use rare gases as the lasing medium. It is also referred to as a krypton ion laser. The stimulated emission process in ion lasers occurs between the two energy states of the ion. Argon is one of the most common gases to be used in gas lasers, as it generates white light.

The unique feature of krypton lasers is that with the use of proper mirrors it will lase on four sharp spectral lines of red, blue, yellow and green colour. The krypton laser is capable of emitting lights of a number of wavelengths (as many as 10), the most significant one being the ones in the visible spectrum of the electromagnetic spectrum.

Krypton lasers are similar to the argon laser in terms of construction and emitted energy levels. The main differences in construction are in the gases used for filling the plasma tube and the mirrors for the desired outputs. The efficiency of this laser is fairly low, due to the high input energy required to ionise the atoms and excite them to the proper state.

Krypton lasers are chiefly used in medical applications, holography and entertainment. Like argon lasers one needs to exercise caution while operating this laser. Proper eye protection should be worn while working with krypton lasers.

Laser Properties	
Laser type	Gas
Pump source	Electric Discharge

Operating Wavelengths
406.7 nm
413.1 nm
415.4 nm
468.0 nm
476.2 nm
482.5 nm
520.8 nm

530.9 nm
568.2 nm
647.1 nm
676.4 nm

Applications

Like argon lasers, krypton ion lasers are primarily used in a number of medical treatments. Krypton lasers are used in various ophthalmic procedures and for the coagulation of the retina. Some of the other applications of this laser are in:

- Forensic medicine.

- Holography and optical pumping source.

- Spectroscopy and microscopy.

- Laser shows in entertainment.

- High speed printing, copying and typesetting.

- Electro-optics research.

The Free Electron Laser

A free electron laser is a relatively exotic type of laser where the optical amplification is achieved in an undulator, fed with high energy (relativistic) electrons from an electron accelerator. Such devices have been demonstrated with emission wavelengths reaching from the terahertz region via the far- and near-infrared, the visible and ultraviolet range to the X-ray region, even though no single device can span this whole wavelength range.

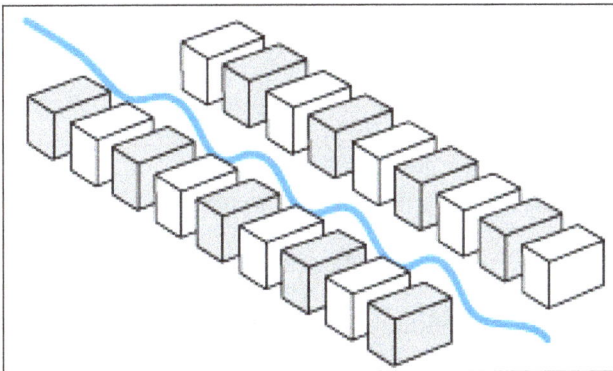

Setup of an undulator, as used in a free electron laser. The periodically varying magnetic field forces the electron beam (blue) on a slightly oscillatory path, which leads to emission of radiation.

In the undulator, a periodic arrangement of magnets (permanent magnets or electromagnets) generates a periodically varying Lorentz force, which forces the electrons to radiate with a frequency which depends on the electron energy, the undulator period, and (weakly) on the magnetic field strength. Both spontaneous and stimulated emission occur allowing for optical amplification in a certain wavelength range.

The greatest attractions of free electron lasers are:

- Their ability to be operated in very wide wavelength regions.

- The large wavelength tuning range possible with a single device.

- The spectacular performance in extreme wavelength regions, not reachable with any other light source.

Compared with other synchrotron radiation sources (pure undulators and wigglers), FELs can generate an output with a much higher spectral brightness and coherence. This is very useful for a number of applications, including fields such as atomic and molecular physics, ultrafast X-ray science, advanced material studies, ultrafast chemical dynamics, biology and medicine.

The big drawback of FELs is that their setups are very large and expensive, so that they can be used only at relatively few large facilities in the world.

High Power Lasers

Lasers with high output powers are required for a number of applications, e.g. for:

- Material processing (welding, cutting, drilling, soldering, marking, surface modification).

- Large-scale laser displays ($\rightarrow$ RGB sources).

- Remote sensing (e.g. with Lidar).

- Medical applications (e.g. Surgery).

- Military applications (e.g. Anti-missile weapons).

- Fundamental science (e.g. Particle acceleration).

- Laser-induced nuclear fusion (e.g. In the NIF project).

Material processing with high-power lasers is the second largest segment of laser applications concerning global turnovers (after communications).

There is no commonly accepted definition of the property "high power"; in the context of laser material processing, it usually means multiple kilowatts or at least a few hundred watts, whereas for laser displays some tens of watts many already be considered high. In some areas, this label is assigned simply for generating a significantly higher output power than other lasers based on the same technology; for example, some "high-powered" laser pointers emit a few hundred milliwatts, whereas ordinary laser pointers are limited to a few milliwatts.

Additional aspects come into play for pulsed lasers. For example, the peak power may be as important as the average output power for a Q-switched laser. Depending on the pulse repetition rate and pulse duration, the peak power may be very high even for a laser with a moderate average output power. Usually, a high average power and not only a high peak power is expected from a high-power laser.

Technical Challenges

The generation of high optical powers in lasers involves a number of technical challenges:

- One requires one or several powerful pump sources. While lamp pumping was originally the only viable approach for most solid-state lasers, pumping with high-power laser diodes (diode bars or diode stacks) has become more and more widespread. Diode-pumped lasers now offer the highest output powers in continuous-wave operation. For very high pulse energies (e.g. tens of joules), lamp pumping is still more practical.

- At least for long-term continuous-wave operation, a high wall-plug efficiency is an important economic factor. Unfortunately, various technical challenges (e.g. thermal effects) tend to make it more difficult at very high power levels to achieve a good efficiency.

- Even in a fairly efficient gain medium, a substantial fraction of the pump power is converted into heat, which can have a number of detrimental side effects. In the worst case, thermally induced stress leads to fracture of the laser crystal. High-power solid-state lasers also exhibit strong thermal lensing, making it substantially more difficult to achieve a high beam quality. In lasers with polarized output, depolarization loss often compromises the efficiency. Efficient heat removal and thermal management are therefore important issues, and additional measures (e.g. in the context of resonator design) are often required for coping with various kinds of thermal effects.

- Particularly in Q-switched lasers, very high optical intensities can occur, which may lead to laser-induced damage of optics (such as laser mirrors) e.g. via laser-induced breakdown. Even if the optical intensities remain well below the damage threshold of all optical elements, tiny dust particles can provoke damage phenomena. It can therefore be essential to keep the laser setup very clean, e.g. by operating it in a sealed case which may be opened only in a clean room. In addition, it can be imperative to use precision optics with a high optical damage threshold.

- Various types of nonlinear effects can also become relevant, particularly in high-power fiber lasers. Examples are stimulated Raman scattering, Brillouin scattering and four-wave mixing.

- Laser resonators with large effective mode areas tend to be sensitive to misalignment and vibrations of optical components. It can therefore be more challenging to achieve robust maintenance-free operation and a good beam pointing stability.

Types of High-power Lasers

There are several different types of high-power lasers:

- High-power diode bars and diode stacks have already been mentioned above as possible pump sources for solid-state lasers. They allow the generation of kilowatts of output power, but with a poor beam quality. For some applications, where beam quality is not essential, the direct use of high-power laser diodes ($\rightarrow$ direct diode lasers) e.g. for laser welding, soldering and brazing, cladding and heat treatment, is an interesting option, offering a comparatively simple, compact, cost-effective and energy-efficient solution.

- There are various types of lamp-pumped or diode-pumped solid-state bulk lasers. Rod lasers can be optimized for several kilowatts of output power, but diffraction-limited beam quality is possible only up to a few hundred watts (with significant efforts). Slab lasers can be developed for tens of kilowatts or more with relatively high beam quality. Thin-disk lasers easily generate hundreds of watts with diffraction-limited beam quality and have the potential to reach that even at power levels well above 10 kW. The power efficiency is usually fairly good.

- High-power fiber lasers and amplifiers can generate up to a few kilowatts with close to diffraction-limited beams and high power efficiency. With relaxed beam quality requirements, even significantly higher powers are possible. Strictly, such fiber devices are often not lasers, but master oscillator power amplifier (MOPA) configurations.

- Some gas lasers, e.g. CO_2 lasers and excimer lasers, are also suitable for hundreds or thousands of watts of output power. They typically operate in different other regions than solid-state lasers, e.g. in the mid-infrared or ultraviolet region.

- There are chemical lasers with multi-kilowatt or even megawatt output powers, explored e.g. in the context of anti-missile weapons.

A perhaps not very practical, but theoretically very interesting high-power laser concept is that of the radiation-balanced laser. Here, the heat generation in the gain medium is essentially eliminated by optical refrigeration.

An aspect of great importance for further laser development is that of power scaling, based on certain power-scalable laser architectures. Even for not power-scalable laser types, it can be very helpful to understand the scaling properties of various parts or techniques.

Safety Issues

The use of high-power lasers raises important issues on laser safety:

- The output powers are far higher than what any eye can tolerate, so that even tiny parasitic reflections must be safely prevented from reaching an eye. Even the use of strongly absorbing laser goggles may not be sufficient as such glasses may not be able to stand such high optical intensities for more than a brief moment of time.

- The skin and clothes of workers are also at risk in environments where optical powers and intensities are sufficient e.g. for laser cutting of metals.

- High-power laser beams may incinerate materials such as plastics or wood. That happens easily already for laser powers of the order of 1 W. Fire protection is therefore an important issue. Also, the formation of poisonous fumes needs to be avoided, or such fumes have to be efficiently removed.

- There are various types of risks which are not related to the laser beams themselves. In particular, high-power electric power supplies often involve high electric voltages, which can cause electric shocks. Power cables, which can be damaged in a harsh industrial environment, can also create hazards.

An important safety principle in the area of high-power lasers is to enclose the laser set-up with a solid housing, and ideally also the whole area where dangerous laser beams can be present. Interlocks can prevent the operation of a laser at times where persons are in a hazardous area.

Tunable Laser

A tunable laser (alternative spelling: tuneable laser) is a laser the output wavelength of which can be tuned (i.e. adjusted) ($\rightarrow$ wavelength tuning). In some cases, a particularly wide tuning range is desired, i.e. a wide range of accessible wavelengths, whereas in other cases it is sufficient that the laser wavelength can be tuned (factory-set) to a certain value. Some single-frequency lasers can be continuously tuned over a certain range, whereas others can access only discrete wavelengths or at least exhibit mode hops when being tuned over a larger range. Lasers are sometimes called wavelength agile or frequency agile when the tuning can be done with high speed.

Tunable lasers are usually operating in a continuous fashion with a small emission bandwidth, although some Q-switched and mode-locked lasers can also be wavelength tuned. In the latter case, it is possible to shift either the envelope of the frequency comb or the lines in the optical spectrum.

There are also other kinds of wavelength-tunable light sources, which often allow tuning over even larger wavelength ranges and are tentatively less costly than tunable lasers. However, they are typically much more limited in terms of radiance and particularly spectral radiance.

Widely Tunable Lasers

Some types of lasers offer particularly broad wavelength tuning ranges:

- A few solid-state bulk lasers, in particular titanium–sapphire lasers and Cr:ZnSe and Cr:ZnS lasers allow tuning over hundreds of nanometers in the near and mid-infrared spectral region. In general, transition-metal doped gain media offer larger tuning ranges than rare-earth-doped gain media, since the electrons involved in such media interact more strongly with the host lattice. Output powers can be hundreds or even thousands of milliwatts.

- Dye lasers also allow for broadband tunability. Different dyes can cover very broad wavelength ranges, e.g. throughout the visible region. There are narrow-linewidth dye laser systems (continuous-wave or pulsed) for use in laser spectroscopy, and also mode-locked dye lasers generating femtosecond pulses.

- Some free electron lasers can cover enormously broad wavelength ranges, and often in extreme spectral regions.

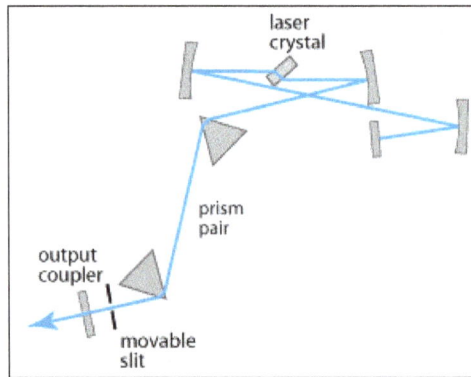

Figure: Setup of a tunable solid-state bulk laser, realized e.g. with a Ti:sapphire laser crystal. The prism pair spatially disperses the different wavelength components, so that the movable slit can be used to shift the wavelength away from that of maximum gain.

Other types of lasers offer tuning ranges spanning a few nanometers to some tens of nanometers:

- Rare-earth-doped fiber lasers, e.g. based on ytterbium, can often be tuned over tens of nanometers, sometimes even more than 100 nm. Most Raman fiber lasers also have the potential for wideband tuning.

- Some rare-earth-doped laser crystals, often doped with ytterbium, also allow for substantial tuning ranges of bulk lasers. Examples are tungstates, vanadates, Yb:BOYS, and Yb:CALGO.

- Color center lasers rely on broadband gain from certain lattice defects in a crystal, which can be generated e.g. with gamma irradiation. They are not widely used, however.

- Most laser diodes can be tuned over a few nanometers (often by varying the junction temperature), but some special types such as external-cavity diode lasers and distributed Bragg reflector lasers can be tuned over 40 nm and more.

- Quantum cascade lasers are also broadly tunable mid-infrared laser sources.

Some fine tuning, often continuously without mode hops, is possible for other lasers:

- Some compact solid-state bulk lasers such as nonplanar ring oscillators (NPROs, MISERs) allow continuous tuning within their free spectral range of several gigahertz. Tuning may be accomplished by applying stress to the laser crystal via a piezo, or by varying the crystal temperature.

- Similar fine tuning is possible with some single-frequency laser diodes, e.g. by varying the drive current.

For wideband tuning in various spectral regions, optical parametric oscillators (OPOs) can be used. These are actually not lasers, but OPO sources are nevertheless sometimes included with the term tunable laser sources.

Wavelength-swept Lasers

There are certain Juniper lasers which are optimized such that the output wavelength can be periodically and rapidly swept through a substantial range. They are called wavelength-swept lasers .

Applications of Tunable Lasers

Wavelength-tunable laser sources have many applications, some examples of which are:

- In laser absorption spectroscopy, a wavelength-tunable laser with narrow optical bandwidth can be used for recording absorption spectra with very high frequency resolution. In a LIDAR system, a laser may be tuned to a wavelength which is specific to a certain substance to be monitored.

- Various methods of laser cooling require a laser wavelength to be adjusted very precisely at or near some atomic resonance.

- Tuning to atomic resonances is also used in laser isotope separation. The laser is then tuned to a particular isotope in order to ionize these atoms and subsequently deflect them with an electric field.

- A tunable laser can be used for device characterization, e.g. of photonic integrated circuits.

- In optical fiber communications with wavelength division multiplexing, a tunable laser can serve as a spare in the case that one of the fixed-wavelength lasers for the particular channels fails. Even though the cost for a tunable laser is higher, its use can be economical as a single spare laser can work on any transmission channel where it is needed. As the cost of tunable lasers is no longer much higher than for non-tunable ones, tunable lasers are now often even used throughout.

- In optical frequency metrology, it is often necessary to stabilize the wavelength of a laser to a certain reference standard (e.g. a multipass gas cell or an optical reference cavity). This can be accomplished e.g. with an electronic feedback system, which automatically adjusts the laser wavelength.

- Some interferometers and fiber-optic sensors profit from a wavelength-tunable laser source, e.g. if this makes it possible to remove an ambiguity or to avoid mechanical scanning of an optical path length.

References

- Introduction-to-solidstate-lasers, lasers-sources: laserfocusworld.com, Retrieved 29 March, 2019

- Rubylaserdefinitionconstructionworking, laser, physics: physics-and-radio-electronics.com, Retrieved 5 February, 2019

- Ndyaglaser, laser, physics: physics-and-radio-electronics.com, Retrieved 25 February, 2019

- Ndyaglaser, laser, physics: physics-and-radio-electronics.com, Retrieved 16 January, 2019

- How-fiber-lasers-work, industrial-fiber-lasers: spilasers.com, Retrieved 19 April, 2019

- Dye-laser, microwave-radar, electronics: daenotes.com, Retrieved 17 May, 2019

- Semiconductor-laser, microwave-radar, electronics: daenotes.com, Retrieved 14 July, 2019

- Heliumneonlaser, laser, physics: physics-and-radio-electronics.com, Retrieved 31 March, 2019

- Nitrogen-lasers, lasers, optical-components-optics, learnmore: globalspec.com, Retrieved 8 August , 2019

- Faq-what-is-an-argon-ion-laser, faqs, technical-knowledge: twi-global.com, Retrieved 9 May, 2019

- Free-electron-lasers: rp-photonics.com, Retrieved 1 August , 2019

3

Laser Characteristics

Some of the diverse characteristics which are studied regarding lasers are the optical cavity, line width, beam quality, optical intensity, peak power and slope efficiency. The topics elaborated in this chapter will help in gaining a better perspective about these characteristics of lasers.

The Optical Cavity

An optical cavity, or an optical resonator, may be described as an arrangement of mirrors that produce a standing light wave resonator. Optical cavities, along with their optical gain, are an integral part of lasers, optical parametric oscillators, and light interferometers.

The simplest form of optical cavity consists of two or more mirrors that are arranged in such a way that light is made to propagate in a closed path. The study of an optical cavity, or resonator, is a vast field, hence is treated as a separate optical subject for study. The standing wave patterns produced in an optical cavity are called modes.

Types of Optical Cavity and Modes

There are two basic types of optical resonator modes. These are: the standing wave resonator, and the travelling wave resonator. The travelling wave resonator is also referred to as the ring resonator.

- Longitudinal or Standing Wave Resonator: The light waves in this type of optical cavity differ in their frequencies.

- Travelling, or Ring, Wave Resonator: The light waves produced in this type of optical cavity differ on both frequency and the intensity pattern.

Common types of resonators are flat and spherical mirror resonators. These types differ mainly in the focal lengths and the distance between the mirrors:

- Flat or Plane-Parallel Optical Cavity: This type of optical cavity consists of two opposing flat mirrors. This is also referred to as the Fabry-Perot Cavity. Despite the simplicity of this optical cavity, it is not used for large-scale laser systems, due to the difficulty in alignment.

- Spherical Cavity: This type of optical cavity exhibits optical resonance when the size of the sphere, refractive index, or the optical wavelength is varied.

Design of Optical Cavities

Flat mirrors are not preferred in building optical cavities. This is due to the difficulty in aligning them. The geometry of the cavity must be such that the light beam remains stable. Some other factors that are considered while designing an optical cavity are Q factor and beam width.

Applications of Optical Cavities

Some of the application areas of optical cavities are listed below:

- Laser resonators;

- Multipass optical delay lines;

- Optical interferometers;

- Laser oscillators.

A laser generally requires a laser resonator (or laser cavity), in which the laser radiation can circulate and pass a gain medium which compensates the optical losses. Exceptions are a few cases where a medium with very high gain is used, so that amplified spontaneous emission extracts significant power in a single pass through the gain medium.

A laser resonator typically contains multiple laser mirrors, one of them being an output coupler, a laser gain medium, and possibly additional optical elements e.g. for wavelength tuning, Q switching or mode locking. It can be a linear resonator, having two end mirrors, or a ring resonator.

The laser radiation is automatically generated at one or several frequencies corresponding to resonances (resonator modes), possibly with small deviations caused by "gain pulling". No special measures are required for operating on the resonance; this is different for external resonators, e.g. resonant enhancement cavities.

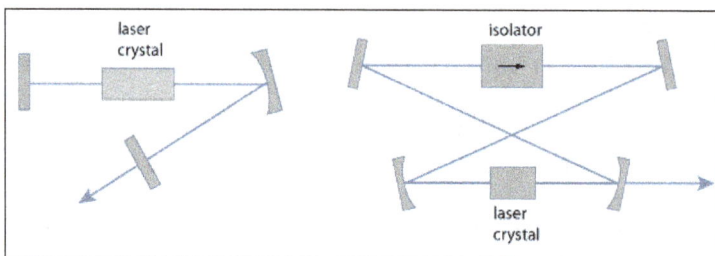

Two simple solid-state laser resonators with a laser crystal as gain medium. Output beams are generated where resonator mirrors are partially transmissive. For the ring laser (right), unidirectional operation is enforced with a Faraday isolator; without that, one would obtain two output beams.

Laser Resonators of Solid-state Lasers

Solid-state bulk lasers are usually built with several dielectric mirrors (laser mirrors), which may be plain or curved. Figure shows a linear resonator and a ring resonator built in that way, and containing a laser crystal as the gain medium. In some cases, a dielectric mirror coating is placed on the gain medium itself. One of the mirrors, usually an end mirror, is the partially transmissive output coupler.

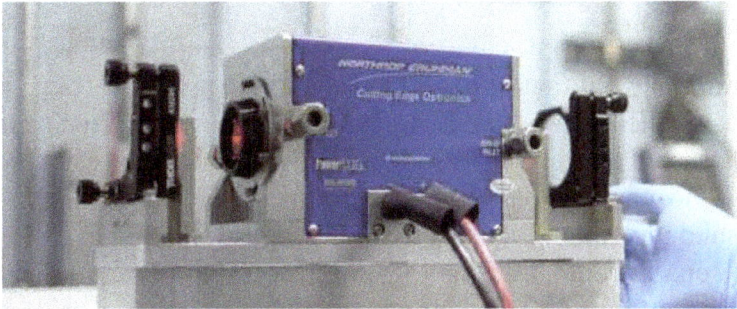

A simple laser resonator consists only of
two mirrors around a diode-pumped laser head.

The design of the laser resonator (comprising optical elements, angles of incidence, and distances between the components) determines the beam radius of the fundamental mode at all locations along the beam, together with other important properties. For maximum beam quality (→ diffraction-limited output), the beam radius in the gain medium has to match approximately the radius of the pumped region. For smaller beam radii, operation with multiple spatial modes is obtained, leading to a non-ideal beam quality; however, such multimode lasers have other advantages such as much wider stability zones and a lower sensitivity to misalignment.

In many cases, the laser resonator design should have additional features. For example, it can be optimized:

- For compactness.

- For achieving certain values of the beam radius in other optical components (e.g. On a saturable absorber of a passively mode-locked laser).

- For avoiding too small beam radii in optical components (leading to optical damage particularly in q-switched lasers).

- For minimizing adverse effects of thermal lensing and related optical aberrations in the gain medium.

- For minimizing the alignment sensitivity.

- For accommodating multiple laser heads.

- For a certain resonator length, determining the pulse repetition rate in a mode-locked laser or the pulse duration of a q-switched laser.

Particularly for high-power lasers with good beam quality, thermal lensing in the gain medium is very important. The resonator design should be made so that changes of the thermal lens do not affect too much the mode sizes. Also, it should have a low sensitivity to thermal aberrations and misalignment. The importance of these factors should not be underestimated; there are cases where two resonators even with equal mode sizes in the gain medium lead to very different laser performance and are radically different in terms of alignment.

Although it is normally not that difficult to evaluate the properties of a given laser resonator, it can be challenging to find a resonator design which satisfies multiple criteria such as those listed above. Numerical optimization, using special resonator design software, can be the only way to find good solutions, particularly for some mode-locked lasers. Also, a solid understanding of resonator properties can help considerably when trying to find resonator configurations with special combinations of properties, such as large mode areas and short lengths. For advanced design issues, a great deal of experience is at least as important as a versatile design software.

Some high-power lasers (for example with slab designs) are operated with unstable resonators, allowing a reasonable (but typically not diffraction-limited) beam quality to be achieved despite the presence of strong thermal effects in the gain medium. Due to the high diffraction losses, such laser cavities require relatively high gain.

There are various types of monolithic solid-state lasers which have the whole beam path within the laser crystal. Beam reflections are then typically realized either with dielectric coatings on crystal surfaces, or with total internal reflection.

Physical Limitations

Although various properties of laser resonators can be optimized with a suitable resonator design, there are limitations, particularly for certain combinations of properties. For example, one can only to a limited extent combine the features of a short resonator length, large mode areas and low alignment sensitivity. Even optimized resonator designs cannot fully meet desirable specifications for certain lasers, particularly high-power lasers.

Alignment of Laser Resonators

For laser resonators with simple designs, e.g. with just two mirrors around some gain medium, the initial alignment is often quite easy to find. Once the laser works, the alignment can be further optimized, simply maximizing the output power.

For more complicated resonators, it can be quite challenging to find some approximate initial alignment where the laser starts operating. In such cases, one may require some visible alignment laser, which should preferably have an appropriate wavelength, such that the laser mirrors have a high enough reflectivity for that beam.

Particularly laser resonators with large mode sizes can have a high alignment sensitivity. Even small tilts of laser mirrors, for example, may move the resonator mode such that the output power drops and possibly the beam quality is degraded.

Common Cavity Configuration

Consider the case of a resonator with two plane mirrors. Parameters g_1 and g_2 are both equal to unity (1), so the arrangement is stable although stability is marginal (ie, the product of $g_1 g_2$ is 1). In practical terms, marginally stable means extreme difficulty in alignment, and a cavity that can become misaligned very easily, usually resulting in the ceasing of lasing oscillation. For this reason, two plane mirrors are rarely used.

For any cavity resonator consisting of two spherical mirrors, the arrangement is stable within limits, as determined by the g parameters. A true confocal arrangement in which the radius of both mirrors is exactly equal to the separation between the mirrors (ie, L) has g parameters equal to $g_1 = g_2 = 0$. The product of the g parameters is hence zero again, so this arrangement is stable. The confocal configuration yields the smallest average spot size of any stable resonator with the beam waist being $w^2_0 = L(\lambda/2\pi)$ occurring at the center of the resonator and the largest spot size $w^2_1 = L(\lambda/\pi)$ occurring at each mirror. This defines the diameter of the output beam if collimated (by a lens) at that point.

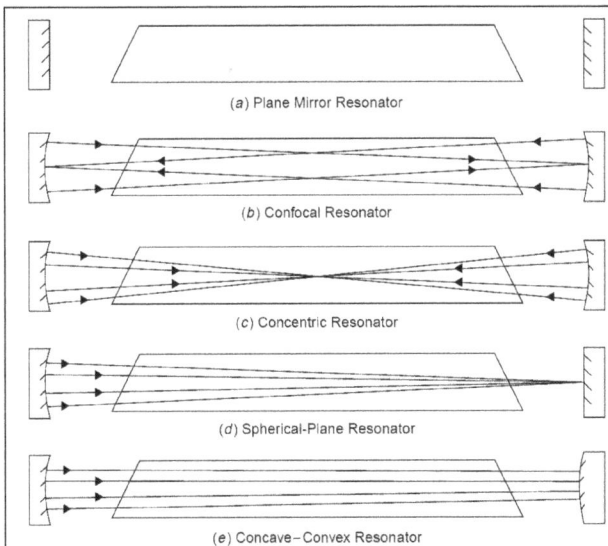

(a) Plane Mirror Resonator

(b) Confocal Resonator

(c) Concentric Resonator

(d) Spherical-Plane Resonator

(e) Concave–Convex Resonator

Common cavity configurations.

Other resonator configurations may have smaller spot sizes at one mirror or the other.

Example: Spot Sizes While a small spot size is desirable for many applications, confocal arrangements do not make efficient use of a large gain volume (such as in CO_2 lasers, which frequently feature a large plasma tube bore). As an example, consider a CO_2 laser with a 1-m-long tube. The spot size at the waist and at each mirror would be:

$$\text{Beam waist:} \quad 2w_0 = \left(\frac{L\lambda}{2\pi}\right)^{1/2} = 2.6\,\text{mm}$$

$$\text{Exit beam:} \quad 2w_1 = \left(\frac{L\lambda}{\pi}\right)^{1/2} = 3.7\,\text{mm}$$

Considering that many CO_2 lasers have plasma tube diameters of between 10 and 25 mm, it is easy to see how inefficient this cavity would be at utilizing the large amplifier volume. In contrast to this, consider a 30-cm-long HeNe in which the spot size at the mirrors would be 0.5 mm. This configuration is considerably more reasonable for a laser of this type since many HeNe tubes have plasma tubes with a diameter of 1 mm.

The confocal arrangement is extremely tolerant to misalignment of either mirror. A small angular tilt of either mirror still maintains the center of curvature of one mirror on the surface of the second cavity mirror, as shown in figure. It is also forgiving of manufacturing tolerances in the radius of the cavity mirrors. Because of this feature, it is an excellent choice for a research laser, where frequent alignment may be required or where alignment cannot be performed by "rocking" or other means. Figure is a concentric configuration in which the radius of each mirror is exactly L/2. It represents a confocal resonator, where the radius of curvature is reduced to its lowest limit. From the point of view of stability, this cavity is stable unless the radius of curvature of the mirrors is even slightly under L/2, in which case the arrangement becomes unstable (this may occur due to alignment or simply, manufacturing tolerances). This arrangement also suffers from difficulties in alignment since the focus of each mirror is coincident and hence difficult to align precisely.

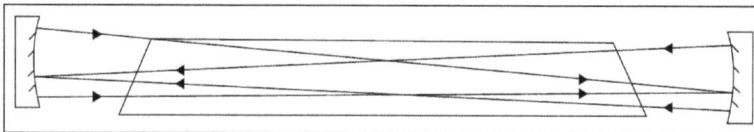

Misalignment of a confocal cavity.

At the upper limit of the radius of curvature the mirrors become plane (and this arrangement is only marginally stable). However, in many practical lasers a variation of a confocal arrangement is used in which mirrors with radii just slightly over L (ie, just longer than the cavity length) are employed. A cavity of this type features a larger waist diameter and hence better utilization of large tube bores. Compared to the true confocal arrangement, these cavities are more tedious to align and so, practically speaking, the radii of curvature are often designed to be only slightly greater (perhaps 5%) than the cavity length.

Other arrangements, such as figure, use a combination of a long-radius spherical mirror (with the radius at least equal to and sometimes much longer than the cavity length) and a plane mirror. This arrangement is the most popular for low and medium power lasers such as HeNe and argon lasers. It is used in many commercial gas lasers (eg, large-frame argons) where the OC is spherical and the HR plane allows the use of various optical configurations. While the OC stays in place, the rear optic may be changed to a wavelength selector for single-line use or a broadband reflector for multiline use.

If the radius of curvature is exactly L (the cavity length), the spot size at the plane mirror is minimal, since this is indeed the beam waist (but again this leads to stability problems similar to those with a true confocal cavity). The beam occupies a cone shaped volume inside the gain medium and so utilizes the amplifier volume more efficiently at one end than the other. In a small HeNe laser, the OC is often flat and is placed at the end of the tube away from the discharge. By placing the active gain medium (which in a small HeNe laser is the actual small capillary tube inside the larger laser tube) near the spherical mirror, the gain volume is used more effectively and higher powers are extracted from it, as shown in figure.

An interesting feature of a spherical-plane cavity configuration is demonstrated in older HeNe tubes, which exploited this feature to assist in alignment of the cavity. During the 1970s and early 1980s, HeNe tubes were made entirely of glass, with mirrors affixed to the ground tube ends by epoxy (called a soft seal, this is no longer used in mass-produced HeNe tubes because such seals allow helium to diffuse through them slowly, and hence the resulting tubes had short lives). The use of a concave mirror allows alignment of the laser cavity using translation of the concave mirror alone; no angular adjustment was required. Figure shows a concave mirror terminating the flat end of a laser which is not perfectly perpendicular to the bore (indeed, it would be quite difficult to grind both glass tube ends perfectly parallel and perpendicular to the bore as would be required for lasing action). This technique of alignment, in which the edges of the mirror were in contact with the flat end of the glass tube, was a great boon to mass manufacturing of these tubes because it required only that the mirror be moved side to side and up / down until alignment was achieved, at which point the mirror was simply fixed in place with epoxy adhesive. Figure shows one of these mirrors, which appears to be haphazardly affixed to the flat end of the glass tube. It is clearly not concentric with the tube bore, but this is exactly how alignment was achieved. The alternative, used on some early HeNe tubes, would be to have each mirror mounted on a three point mount allowing angular adjustment and attaching each mount to the tube using flexible metal tubing.

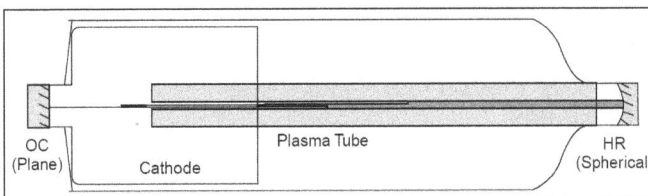

Spherical-plane resonator for a HeNe laser.

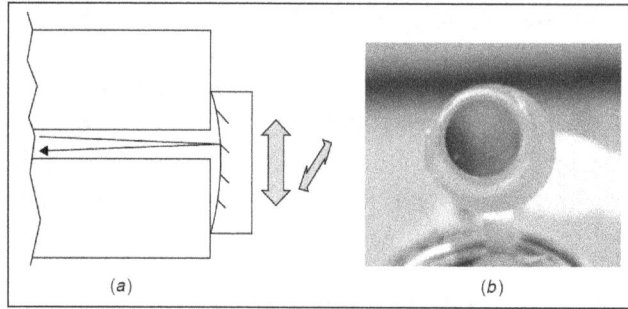

Spherical reflector alignment on a HeNe tube.

In the final resonator arrangement, a concave and a convex mirror are used. Again, this arrangement is stable within the confines set out by the g-parameter equation. Concave – convex resonators can utilize much more of the lasing volume since the smallest spot size, at the focus of the concave mirror, is outside the cavity itself. These types of cavities are very sensitive to misalignment, though, so are rarely used in commercial lasers.

Not all lasers use stable resonators, and for certain high-power lasers such as excimer and carbon dioxide TEA lasers, unstable resonators are a popular option. Consider the two unstable resonator configurations depicted in figure Part (a), a positive branch confocal resonator, consists of a small convex and large concave mirror. The beam exits around the edges of the smaller mirror and has an annular shape. In part (b), a negative branch confocal resonator, the beam is also annular in shape. Because these resonators are not stable, light is not trapped in the cavity, at least for many round trips, so this arrangement is suitable only for use with high gain lasers, in which only a few transits through the gain medium are required to amplify the oscillations to a usable power level. The primary benefit of an unstable resonator is allowing the use of total reflectors that would not be damaged by high power levels (in the tens of kilowatts level) as a partially reflecting output coupler might be.

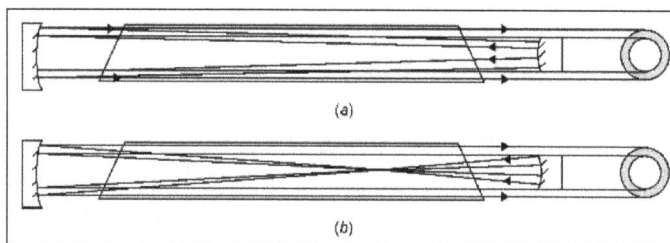

Unstable resonators.

The output beam does not pass through the output coupler itself as it does in a stable resonator configuration. In the case of large carbon dioxide lasers, solid metal mirrors (usually, copper with gold plating and often water cooled) may be used, which have very high damage thresholds.

Examining figure, it is also evident that an unstable resonator utilizes a large volume of the lasing medium, allowing efficient extraction of energy. The biggest apparent problem with this configuration is that the shape of the beam is not Gaussian, so cannot be focused to a sharp point. In reality, it can be focused to almost as sharp a point (certainly better than many stable resonators yield when operating in high-order modes), so is quite suitable for materials-processing applications.

Finally, it may be worth noting that an OC can be avoided altogether in a stable cavity configuration by constructing a cavity in which one mirror has a hole in it. This approach, popular with amateur laser constructors for carbon dioxide lasers, usually utilizes a spherical-plane cavity for ease of alignment. A small hole with a diameter about 10% of the diameter of the plasma tube is drilled into the flat mirror and sealed with a window transparent to infrared radiation. In this manner, inexpensive metal film mirrors (usually, glass coated with copper or gold, both of which reflect infrared radiation well) may be used in place of an OC made of germanium or zinc selenide, which are comparatively expensive. The approach works but does not allow good mode performance and so is rarely used in commercial lasers.

Q Factor

The *Q factor* (quality factor) of a resonator is a measure of the strength of the damping of its oscillations, or for the relative linewidth. The term was originally developed for electronic circuits, e.g. LC circuits, and for microwave cavities, but later also became common in the context of optical resonators.

There are actually two different common definitions of the Q factor of a resonator:

- Definition via energy storage: The Q factor is 2π times the ratio of the stored energy to the energy dissipated per oscillation cycle, or equivalently the ratio of the stored energy to the energy dissipated per radian of the oscillation. For a microwave or optical resonator, one oscillation cycle is understood as corresponding to the field oscillation period, not the round-trip period.

- Definition via resonance bandwidth: The Q factor is the ratio of the resonance frequency v_0 and the full width at half-maximum (FWHM) bandwidth δv of the resonance:

$$Q = \frac{v_0}{\delta v}$$

Both definitions are equivalent only in the limit of weakly damped oscillations, i.e. for high Q values. The term is mostly used in that regime.

Q Factor of an Oscillator

The term *Q factor* is sometimes also applied to continuously operating *oscillators*, such as active optical frequency standards. In that case, only the definition via the bandwidth can be used; the bandwidth is then the linewidth of the output signal.

If the oscillator is based on some resonator (which is virtually always the case), the effective Q factor of the oscillator may deviate substantially from the intrinsic Q value of the resonator. Particularly measurements on atomic transitions (such as in a cesium atomic clock) have a limited measurement time, so that the effective linewidth of the reference transition is increased. (This problem can be severe for cesium clocks; cesium fountain clocks represent a significant advance towards longer measurement times.) On the other hand, a carefully stabilized oscillator can have a linewidth which is a tiny fraction of the linewidth of the underlying frequency standard; for cesium atom clocks, the quartz oscillator is often stabilized e.g. to a millionth of the linewidth of the signal from the cesium beam apparatus. Effectively, the good short-term stability of the quartz oscillator is combined with the high accuracy and low long-term drift of the cesium apparatus.

Q Factor of an Optical Resonator

The Q factor of a resonator depends on the optical frequency v_0, the fractional power loss l per round trip, and the round-trip time T_{rt}:

$$Q = v_0 T_{rt} \frac{2\pi}{l}$$

(assuming that $l \ll 1$).

For a resonator consisting of two mirrors with air (or vacuum) in between, the Q factor rises as the resonator length is increased, because this decreases the energy loss per optical cycle. However, extremely high Q values are often achieved not by using very long resonators, but rather by strongly reducing the losses per round trip. For example, very high Q values are achieved with whispering gallery modes of tiny transparent spheres.

Important Relations

The Q factor of a resonator is related to various other quantities:

- The Q factor equals 2π times the exponential decay time of the stored energy times the optical frequency.

- The Q factor equals 2π times the number of oscillation periods required for the stored energy to decay to $1/e$ ($\approx 37\%$) of its initial value.

- The Q factor of an optical resonator equals the finesse times the optical frequency divided by the free spectral range.

High-Q Resonators

One possibility for achieving very high Q values is to use supermirrors with extremely low losses, suitable for ultra-high Q factors of the order of 10^{11}. Also, there are toroidal silica microcavities with dimensions of the order of 100 μm and Q factors well above 10^8, and silica microspheres with whispering gallery resonator modes exhibiting Q factors around 10^{10}.

High-Q optical resonators have various applications in fundamental research (e.g. in quantum optics) and also in telecommunications (e.g. as optical filters for separating WDM channels). Also, high-Q reference cavities are used in frequency metrology, e.g. for optical frequency standards. The Q factor then influences the precision with which the optical frequency of a laser can be stabilized to a cavity resonance.

Q Factor in Laser Physics

When the Q factor of a laser resonator is abruptly increased, an intense laser pulse (giant pulse) can generated. This method is called Q switching. High-Q laser resonators can be used for obtaining laser output with a very narrow linewidth.

Laser Modes

The cross-sectional intensity distribution of the laser radiation field is not homogenous: the intensity is modulated transverse to the propagation direction and decreases steadily at the edges rather than abruptly. This is due to diffraction which always occurs during propagation and is related to the wave-like nature of light.

If one assumes a homogenous intensity distribution on one of the laser mirrors, the intensity distribution will change during propagation from one mirror to the other due to diffraction. After numerous passes back and forth, the intensity distribution takes shape and is reproduced with each pass. These marked intensity distributions represent the intrinsic solution of the optical resonator of which there are, in fact, several. One of these intrinsic solutions is, however, of particular importance: the so-called fundamental mode. In many cases, the fundamental mode approximates the so-called Gaussian beam. The Gaussian beam has a radial intensity distribution with a Gaussian profile.

A significant parameter involved in designing optical resonators is the Fresnel number which is defined as the radius of the resonator mirror squared divided by the product of the wavelength of the laser radiation and the distance to the laser mirror. Fundamental mode radiation is associated with Fresnel numbers of around 1. Diffraction losses increase with smaller Fresnel numbers; at larger Fresnel numbers, the radiation field modes arise which causes a decline in beam quality. The beam quality determines, for instance, how small the beam cross-section in the focal plane of a lens can be - a crucial factor for many applications. In the fundamental operating mode, the minimal focal radius is solely limited by diffraction - it physically cannot get any smaller.

The effect of diffraction on the transverse beam distribution can be simulated in this applet. For simplicity, the diffraction problem will be considered one-dimensionally based on a strip resonator. The middle section corresponds to the mirror and the radiation intensity to the left and right of the mirror is not re-reflected and is lost. This is known as diffraction losses. The red line represents the intensity of the n-1th pass; the black line represents the nth pass applied over the coordinates transverse to the propagation direction.

Longitudinal Modes

The output frequencies of a laser are determined by several factors. First, the gross wavelength is determined by the energy uncertainty (broadening) of the laser transition, which determines the wavelength and overall linewidth. Nonetheless, at any given instant, only a relatively few frequencies within this overall envelope are allowed to oscillate. These "longitudinal modes" result from the boundary conditions that, in a conventional two-mirror lasers, the amplitude of the wave must be zero at the mirror surface (i.e., that the oscillating wave is a standing wave). This means only those laser frequencies that meet the criteria:

$$n = Nc/2L$$

can operate, where c is the speed of light, L is the effective cavity length, and N is an interger. Adjacent modes are typically orthogonally polarized.

The illustration below shows the lasing envelope of a helium neon laser operating at 632.8nm with a cavity spacing of 23cm. This results in a mode spacing of 640MHz. Since the width of the gain curve (FWHM) is only 1400MHz, only two longitudinal modes can operate at any given time. If the laser were twice as long, four longitudinal modes could operate simultaneously.

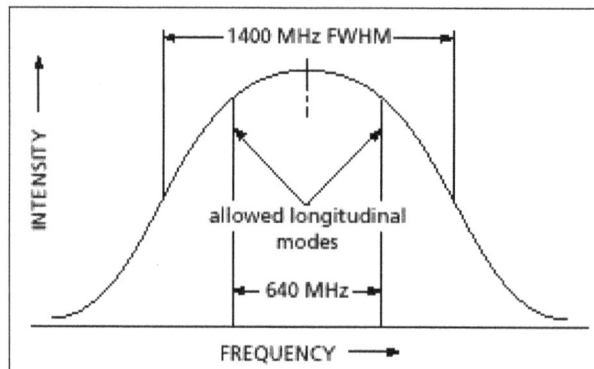

Since the allowable longitudinal modes are a function of cavity length, the frequency will change as the cavity length changes. In lasers where only a few longitudinal modes can operate, these changes will cause outout power to fluctuate as the modes sweep under the gain curve.

Transverse Modes

Transverse modes have a field vector normal to the direction of propogation and are determined by the geometry of the laser or waveguide cavity and any limiting apertures. Waveguide modes are characterized as TE (transverse electric) with the electric vector normal and TE (transverse magnetic) with the magnetic vector normal. In general, laser modes that do not have wall boundary conditions are designated TEM (transvers electric magnetic) with both vectors normal to the direction of propogation. The lowest order mode is the Gaussian TEM_{00}. The appearance that higher-order modes take depends upon whether the limiting apertures are circular or rectancular. The three lowest-order modes for a circularly symmetric cavity are shown below.

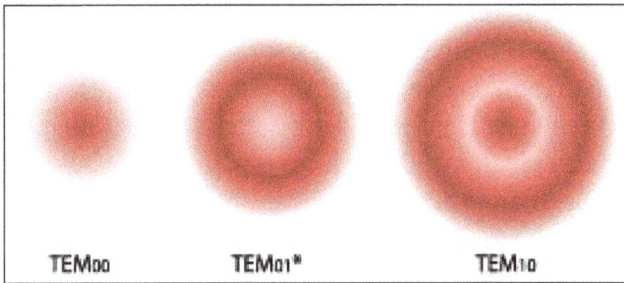

TEM₀₀   TEM₀₁*   TEM₁₀

The higher order the mode, the smaller the beam diameter (for a given geometry), the lower the divergence, and the smaller the M^2 value ($M^2 = 1$ for a pure TEM_{00} beam). The higher order the mode, the more uniform the beam cross-section. Very high-order-mode beams have a top-hat shape.

Line Width of a Laser

The linewidth (or line width) of a laser, e.g. a single-frequency laser, is the width (typically the full width at half-maximum, FWHM) of its optical spectrum. More precisely, it is the width of the power spectral density of the emitted electric field in terms of frequency, wavenumber or wavelength.

Similarly, other spectral lines e.g. from gas discharge lamps have certain linewidths, which can depend on the operation conditions.

The linewidth of a light beam is strongly (but non-trivially) related to the temporal coherence, characterized by the coherence time or coherence length. A finite linewidth arises from phase noise if the optical phase undergoes unbounded drifts, as is the case for free-running laser oscillators, for example. (Phase fluctuations which are restricted to a small interval of phase values lead to a zero linewidth and some noise sidebands.) Drifts of the resonator length can further contribute to the linewidth and can make it dependent on the measurement time. This shows that the linewidth alone, or even the linewidth complemented with a spectral shape (line shape), does by far not provide full information on the spectral purity of laser light. (This is particularly the case for lasers with dominating low-frequency phase noise.) More data are required for full noise specifications.

The r.m.s. linewidth can be defined as the root-mean-square value of the instantaneous optical frequency:

$$\Delta v_{\text{r.m.s}} = \sqrt{\int_{f_1}^{z} S_{\Delta V}\left(f\right)\mathrm{df}}$$

where usually some limited integration range for the noise frequencies is chosen. This quantity can be more easily calculated from the power spectral density $S_{\Delta v}(f)$ of the instantaneous frequency. Note, however, that the r.m.s. linewidth is not always a sensible measure; one should only use it in cases with strongly increasing $S_{\Delta v}(f)$ for decreasing noise frequency (flicker noise), but not e.g. for white frequency noise. The relation between the r.m.s. linewidth and the width of the optical spectrum is not trivial and depends on the shape of the frequency noise spectrum.

Lasers with very narrow linewidth (high degree of monochromaticity) are required for various applications, e.g. as light sources for various kinds of fiber-optic sensors, for laser spectroscopy (e.g. LIDAR), in coherent optical fiber communications, and for test and measurement purposes. Note that the achieved linewidth can be many orders of magnitude below the line with of the used laser transition.

Quantum Noise and Technical Noise

The simplest situation is one where only spontaneous emission (quantum noise) introduces phase noise. In that case, the noise of the instantaneous frequency is white noise, i.e., its power spectral density is constant, and the emission spectrum is of Lorentzian shape. The corresponding linewidth was calculated by Schawlow and Townes even before the first laser was experimentally demonstrated. According to the modified Schawlow–Townes equation (with a correction from M. Lax).

$$\Delta v_{laser} = \frac{\pi h v \Delta\left(v_o\right)^2}{P_{\text{out}}}$$

the linewidth (FWHM) is proportional to the square of the resonator bandwidth divided by the output power (assuming that there are no parasitic resonator losses).

The Schawlow–Townes limit is usually difficult to reach in reality, as there are various technical noise sources (e.g. mechanical vibrations, temperature fluctuations, and pump power fluctuations) which are difficult to suppress. There are therefore certain compromises in laser design for narrow linewidth. For example, a long laser resonator leads to a small Schawlow–Townes linewidth, but makes it more difficult to achieve stable single-frequency operation without mode hops, and to get a mechanically stable setup.

Typical measured linewidths of stable free-running single-frequency solid-state lasers (e.g. for a measurement time of 1 s) are a few kilohertz, which is far above the Schawlow–Townes limit. Various sources of technical noise, e.g. fluctuations of the resonator length, the pump power or the temperature of the laser crystal, can be responsible for the increased linewidth.

The linewidths of monolithic semiconductor lasers are often in the megahertz range and are strongly increased above the Schawlow–Townes limit mainly by amplitude-phase coupling, as described with the linewidth enhancement factor. There can also be excess noise from charge carrier fluctuations with a 1/f characteristic of the PSD of the frequency fluctuations. In that case, the measurement time influences the measured linewidth value.

Much smaller linewidths, sometimes even below 1 Hz, can be reached by stabilization of lasers, e.g. using ultrastable reference cavities. Small linewidths are important, e.g. for spectroscopic measurements and for application in fiber-optic sensors.

Measurement of Laser Linewidth

A laser linewidth can be measured with a variety of techniques:

- For large linewidths (e.g. > 10 GHz, as obtained when multiple modes of the laser resonator are oscillating), traditional techniques of optical spectrum analysis, e.g. based on diffraction gratings, are suitable. A high frequency resolution is difficult to obtain in this way.

- Another technique is to convert frequency fluctuations to intensity fluctuations, using an optical frequency discriminator, which can be, e.g., an unbalanced interferometer or a high-finesse reference cavity. Again, the measurement resolution is quite limited.

- For single-frequency lasers, the self-heterodyne technique is often used, which involves recording a beat note between the laser output and a frequency-shifted and delayed version of it.

- For sub-kilohertz linewidths, the ordinary self-heterodyne technique usually becomes impractical, as one would require a very large delay length, but it can be extended by using a recirculating fiber loop with an internal fiber amplifier.

- Very high resolution can also be obtained by recording a beat note between two independent lasers, where either the reference laser has significantly lower noise than the device under test, or both lasers have similar performance. One can retrieve the instantaneous difference frequency e.g. with PLL (phase-locked loop) following the beat signal, or numerically from digitized recordings. This method is conceptually very simple and reliable, but the requirement of a second laser (operating at a nearby optical frequency) can be inconvenient. If linewidth measurements are required in a wide spectral range, a frequency comb source can be very useful.

Note that an optical frequency measurement always needs some kind of frequency (or timing) reference somewhere in the setup. For lasers with narrow linewidth, only an optical reference can give a sufficiently accurate reference. The self-heterodyne technique is a way to derive the frequency reference from the device under test itself by applying a large enough time delay, ideally avoiding any temporal coherence between the original beam and the delayed version. Therefore, long fibers are often used; however, long fibers tend to introduce additional phase noise due to temperature fluctuations and acoustic influences.

Particularly in cases with $1/f$ frequency noise, a linewidth value alone may not be regarded as completely characterizing the phase noise. It may then be better to measure the whole Fourier spectrum of the phase or instantaneous frequency fluctuations and characterize it with a power spectral density; Note also that $1/f$ frequency noise (or other noise spectra with strong low-frequency noise) can cause problems with some measurement techniques.

Minimization of Laser Linewidth

The linewidth of a laser depends strongly on the type of laser. It may be further minimized by optimizing the laser design and suppressing external noise influences as far as possible. The first step should be to determine whether quantum noise or classical noise is dominating, because the required measures can depend very much on this.

The influence of quantum noise (essentially spontaneous emission noise) is small for a laser with high intracavity power, low resonator losses, and a long resonator round-trip time. Classical noise may be introduced via mechanical fluctuations, which can often be kept weaker for a compact short laser resonator, but note that resonator length fluctuations of a certain magnitude have a stronger effect in a shorter resonator. Proper mechanical construction can minimize the coupling of the laser resonator to external vibrations and also minimize effects of thermal drift. There can also be thermal fluctuations in the gain medium, introduced e.g. by a fluctuating pump power. For superior noise performance, various schemes for active stabilization can be employed, but it is often advisable first to use all practical passive methods.

Single-frequency solid-state bulk and fiber lasers can achieve linewidths of a few kilohertz, or sometimes even below 1 kHz. With serious efforts at active stabilization, subhertz linewidths are sometimes achieved. The linewidth of a laser diode is typically in the megahertz region, but it can also be reduced to a few kilohertz, e.g. in external-cavity diode lasers, particularly with optical feedback from a high-finesse reference cavity.

Problems Resulting from a Narrow Linewidth

A narrow linewidth from a laser source is not always desirable:

- A large coherence length implies that interference effects (e.g. due to weak parasitic reflections) can easily spoil the beam profile. In laser projection displays, laser speckle effects can disturb the image quality.

- For transmission of light in passive or active optical fibers, a narrow linewidth can cause problems due to stimulated Brillouin scattering. It is then sometimes necessary to increase the optical linewidth, for example by fast dithering of the instantaneous frequency via current modulation of a laser diode or with an optical modulator.

Beam Quality

The beam quality of a laser beam is an important aspect of laser beam characterization. It can be defined in different ways, but is normally understood as a measure of how tightly a laser beam can be focused under certain conditions (e.g. with a limited beam divergence). The most common ways to quantify the beam quality are:

- The beam parameter product (BPP), i.e., the product of beam radius at the beam waist with the far-field beam divergence angle.

- The M^2 factor, defined as the beam parameter product divided by the corresponding product for a diffraction-limited Gaussian beam with the same wavelength.

- The inverse M^2 factor, which is high (ideally 1) for beams with high beam quality.

A high beam quality implies smooth wavefronts (i.e., strong phase correlation across the beam profile), such that focusing the beam with a lens allows one to obtain a focus where the wavefronts are plane. Scrambled wavefronts make beam focusing more difficult, i.e., the beam divergence for a given spot size is increased.

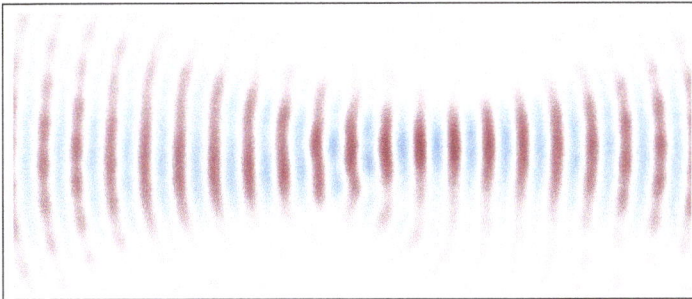

A laser beam with poor beam quality. In contrast to an ideal Gaussian beam, the wavefronts are somewhat scrambled, which makes it more difficult to tightly focus the beam.

The highest possible beam quality in terms of M² is achieved for a diffraction-limited Gaussian beam, having M² = 1. That value is closely approached by many lasers, in particular by solid-state bulk lasers operating on a single transverse mode (→ single-mode operation) and by fiber lasers based on single-mode fibers, also by some low-power laser diodes (particularly VCSELs). On the other hand, in particular some high-power lasers (e.g. solid-state bulk lasers and semiconductor lasers such as diode bars) can have a very large M² of more than 100 or even well above 1000. In solid-state lasers, this is often a result of thermally induced wavefront distortions in the gain medium and/ or a mismatch of effective mode area and pumped area in the laser crystal, whereas in high-power semiconductor lasers the poor beam quality results from operation with a highly multimode waveguide. In both cases, the poor beam quality is associated with the excitation of higher-order resonator modes.

In the focus (beam waist) of a diffraction-limited beam (i.e., at the location where the beam radius reaches its minimum), the optical wavefronts are flat. Any scrambling of the wavefronts, e.g. due to optical components with poor quality, spherical aberrations of lenses, thermal effects in a gain medium, diffraction at apertures, or by parasitic reflections, can spoil the beam quality. For monochromatic beams, the beam quality could in principle be restored e.g. with a phase mask which exactly compensates the wavefront distortions, but this is usually difficult in practice, even in cases where the distortions are stationary. A more flexible approach is to use adaptive optics in combination with a wavefront sensor.

It is possible to some extent to improve the beam quality of a laser beam with a non-resonant mode cleaner or a mode cleaner cavity. This, however, leads to some loss of optical power.

The brightness of a laser is determined by its output power together with its beam quality. The term beam quality is sometimes used with a qualitative meaning which has little to do with the focusability. For some applications, it is vital to obtain a smooth beam intensity profile, e.g. of Gaussian shape, whereas the beam divergence is not of interest. The "quality" of a laser beam may then not be characterized e.g. with an M^2 one beam may have a relatively small M^2 value but a multi-peaked beam profile, whereas another beam may have a smooth beam shape but a high divergence and thus a large M^2 value.

Some laser applications such as lithography require the uniform illumination of a large area. If the term beam quality appears in that context, it may have nothing to do with focusability. One may then even prefer beams with a rather low spatial and temporal coherence.

Measurement of Beam Quality

According to ISO Standard 11146, the beam quality factor M^2 can be calculated with a fitting procedure, applied to the measured evolution of the beam radius along the propagation direction (the so-called caustic). For correct results, a number of rules have to be observed, e.g. concerning the exact definition of the beam radius and the placing of data points.

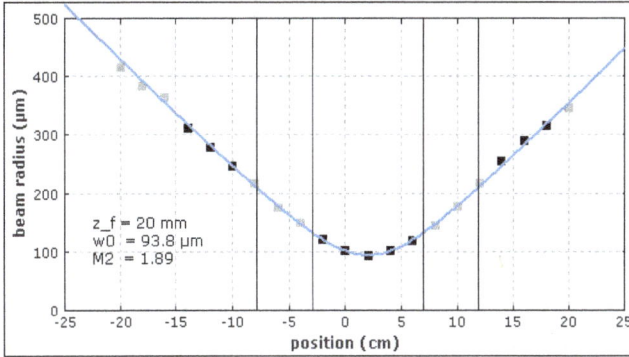

Calculation of the beam quality from the measured caustic. The black data points are those used for the fitting procedure, whereas the gray points are ignored. (A balanced selection of data points, with some near the beam waist and others at a sufficient distance from it, is required according to ISO Standard 1114).

There are commercially available beam profilers which can automatically perform beam quality measurements within a few seconds. They are normally based on the measurement of the beam profile at different positions. Beam profilers based on different measurement principles, e.g. CCD and CMOS cameras or rotating knife edges or slits, differ considerably in terms of the allowed ranges of beam radius and optical power, wavelength range, sensitivity to artifacts, etc. For example, slit or knife-edge scanners can usually handle higher powers than cameras and can be precise for nearly Gaussian-shaped beams, whereas camera-based systems are usually more appropriate for complicated beam shapes. Other issues come into play for beams with temporally varying powers, e.g. for the output of Q-switched lasers. It may then be necessary to synchronize a shutter with the laser pulses. Instead of a moving a detector through the beam, one can use a spatial light modulator to avoid any moving parts.

Alternative measurement methods are based on the transmission through a mode-matched passive optical resonator or on wavefront sensors, e.g. Shack–Hartmann wavefront sensors. The full characterization of the laser beam then only requires analysis in a single plane.

Importance of Beam Quality for Applications

A high beam quality can be important e.g. when strong focusing of a beam is required. In the area of laser material processing, printing, marking, cutting and drilling require high beam qualities, whereas welding and various kinds of surface treatment are less critical in this respect, because they work with larger spots, so that direct application of high-power laser diodes with poor beam quality is possible. For cutting and remote welding, a relatively high beam quality (with M^2 not much larger than 10) makes it possible to use a large working distance (i.e., a large distance between workpiece and focusing objective), which is highly desirable e.g. in order to protect the optics against debris and fumes. Also, a high beam quality reduces the beam diameters in a beam delivery system, so that smaller and thus cheaper optical elements (e.g. mirrors and lenses) can be used. Furthermore, the increased effective Rayleigh length (for a given spot size) increases the tolerance for longitudinal alignment.

A large working distance, made possible by a high beam quality, is also important for the design of diode-pumped lasers when the pump beam has to go through various pieces of optics (e.g. a dichroic mirror) before reaching the laser crystal.

A very high (close to diffraction-limited) beam quality, associated with a high spatial coherence, is often required for interferometers, optical data recording, laser microscopy, and the like.

Mode-locked lasers always have to have a high beam quality, since the excitation of higher-order transverse modes would disturb the pulse formation process.

Typical Beam Quality of Certain Lasers

Generally, the beam quality is not determined by the type of laser, but there are some typical trends:

- Most low-power diode-pumped solid-state lasers exhibit a high (close to diffraction-limited) beam quality.

- The same applies to various gas lasers such as helium–neon lasers and CO_2 lasers.

- Some high-power solid-state lasers exhibit a poor beam quality, essentially because strong thermal effects in the laser crystal lead to beam distortions. Also, there can be design trade-off between high beam quality and high power efficiency, or high beam quality and low alignment sensitivity.

- Low-power laser diodes normally have a rather high beam quality, whereas high-power laser diodes basically always have a poor beam quality. Essentially, this is because high powers require large emitting apertures which make the used waveguides highly multimode. (The numerical aperture cannot be strongly reduced).

Optimizing Laser Beam Quality

Crucial factors for obtaining a high beam quality from a solid-state bulk laser are:

- An optimized resonator design with suitable mode area (particularly in the gain medium) and low sensitivity to thermal lensing.

- Good resonator alignment.

- Minimized thermal effects, particularly from thermal lensing in the gain medium.

- High-quality optical components (particularly concerning the gain medium).

- An optimized pump intensity distribution (sometimes requiring a pump source with good beam quality) – more easily achieved with end pumping than with side pumping.

Beam Quality in Nonlinear Optics

Beam quality is an issue not only for lasers, but also for nonlinear frequency conversion. While thermal lensing in nonlinear crystal materials occurs only at very high average power levels (because heating occurs only through weak parasitic absorption), the beam quality can be affected by other effects:

- Spatial walk-off can spatially shift the interacting beams, so that the overlap becomes weaker, and the interaction becomes spatially asymmetric.

- For strong conversion e.g. in a frequency doubler or an optical parametric amplifier, there can be strong depletion of the pump beam near the beam axis or even backconversion, in extreme cases leading to pronounced ring structures. Gain guiding can make such problems more severe. Beam quality issues have been shown to limit the power scalability of high-gain nonlinear frequency conversion devices.

- For ultrashort pulses, the group velocity mismatch and other effects can even lead to time-dependent beam quality.

Further, the use of a laser beam with poor beam quality in a nonlinear frequency conversion device can significantly spoil the conversion efficiency.

Beam quality effects in nonlinear optics can be investigated with numerical computer models, which can simulate the evolution of the spatial (and possibly temporal) profiles of the involved beams.

Optical Intensity

The optical intensity I, e.g. of a laser beam at some location, is the optical power per unit area, which is transmitted through an imagined surface perpendicular to the propagation direction. The units of the optical intensity (or light intensity) are W/m^2 or (more commonly) W/cm^2. The intensity is the product of photon energy and photon flux. It is sometimes called optical energy flux.

In radiometry and photometry, *intensity* is often understood to be the radiant or luminous power per unit solid angle; this must not be confused with the optical intensity as usually used in optics and laser technology.

When light is received by a surface, an optical intensity causes an irradiance, which is the intensity times the angle against normal direction. The two quantities are different, although they have the same units.

Optical intensities and powers are normally understood as quantities which are averaged over at least one oscillation cycle. In other words, they are *not* considered to be oscillating on the time scale of an optical oscillation.

For a monochromatic propagating wave, such as a plane wave or a Gaussian beam, the local intensity is related to the amplitude E of the electric field via:

$$l = \frac{\upsilon_p \varepsilon_0 \varepsilon_1 \mu_r}{2} |E|^2 = \frac{c \varepsilon_0 n}{2} |E|^2$$

where υ_p is the phase velocity, c is the vacuum velocity of light, and n is the refractive index. For non-monochromatic waves, the intensity contributions of different spectral components can simply be added, if beat notes are not of interest.

The above equation does not hold for arbitrary electromagnetic fields. For example, an evanescent wave may have a finite electrical amplitude while not transferring any power. The intensity should then be defined as the magnitude of the Poynting vector.

For a laser beam with a flat-top intensity profile (i.e., with a constant intensity over some area, and zero intensity outside), the intensity is simply the optical power P divided by the beam area. For a Gaussian beam with optical power P and Gaussian beam radius w, the peak intensity (on the beam axis) is:

$$l_p = \frac{P}{\pi W^2 / 2}$$

which is two times higher than is often assumed. The equation can be verified by integrating the intensity over the whole beam area, which must result in the total power.

In a multimode laser beam, generated in a laser where higher-order transverse resonator modes are excited, the shape of the transverse intensity profile can undergo significant changes as the relative optical phases of the modes change with time. The peak intensity can then vary, and may occur at locations at some distance from the beam axis.

The term intensity is often used in a non-quantitative or not very precise way, and not clearly distinguished from the optical power. For example, the intensity noise normally refers to noise (fluctuations) of the optical power, rather than the intensity.

Optical intensities are relevant in various situations:

- In conjunction with transition cross sections, intensities govern the rates of optical transitions, e.g. in laser gain medium. Strong saturation of an optical transition in the steady state occurs when the intensity exceeds the saturation intensity.

- The refractive index change via the Kerr effect in a transparent medium is the nonlinear index times the local intensity.

- Laser-induced damage of a medium may occur for intensities above a certain damage threshold, which however can usually only be reached with optical pulses, and then depends on the pulse duration.

- Extremely high peak intensities can be achieved with amplified ultrashort pulses. For intensities of e.g. 10^{14} W/cm² or higher in a gas, high harmonic generation can occur.

- The radiation pressure of light incident on a surface is proportional to the optical intensity.

Beam profilers can be used for measuring the shape of the intensity profile of a laser beam.

Peak Power of Laser

Peak power is formally defined as the maximum optical power a laser pulse will attain. In more loosely-defined terms, it is an indicator of the amount of energy a laser pulse contains in comparison to its temporal duration, namely pulse width.

A laser with high peak power is one that has pulses that are either high in energy per pulse or short in pulse width, but generally, both conditions are combined.

For a flat top beam, this is mathematically represented by:

$$Peak\ power\ (W) = \frac{Energy\ per\ pulse\ (J)}{Pulse\ width\ (s)}$$

For a Gaussian beam, one can apply a factor of 2 to represent the actual peak power generated in the middle of the beam. In turn, peak power density is defined as peak power divided by the area covered by the laser spot. Note that peak power is quite different from average power, the latter being loosely-defined as the amount of optical energy a laser produces each second. Peak power has no real significance when it comes to a CW laser.

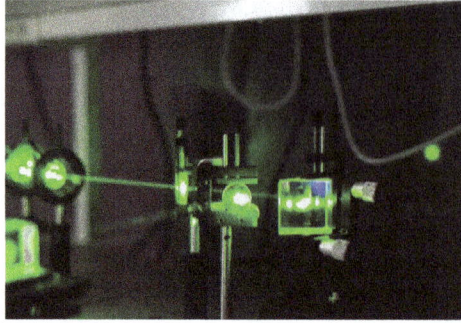

Values for both peak power and peak power density for most lasers are typically immense due to the prevalence of fast laser sources on the market. Even ultrafast lasers are now a thing. Therefore, it is quite possible to hear about a gigawatt laser, or even a petawatt laser, as a reference to the peak power values these lasers can reach.

To be specific, energy density divided by pulse width will return a peak power density value, yet, energy density threshold diminishes as pulse width gets shorter.

Consequently, this implies that lasers with an energy density level that is within the safe range of a laser power meter or a laser energy meter will also be within the safe range in regards to peak power density.

Turning the Spotlight on the Meters

Energy can be both concentrated spatially (energy density) and temporally (peak power), but also both at once (peak power density). It is therefore wise to describe the effects of high peak power density (and therefore energy density) on a detector, along with some precautions for people to consider.

One can imagine that such concentrated energy will literally vaporize the absorber if precautions are not taken. Two simple tips you should consider, if you are having trouble, are:

1.   Expanding the beam diameter to get a larger laser spot hitting the absorber of the detector;

2.   Opt for a more resilient absorber (like our proprietary absorbers W or VR).

Finally, note that some laser parameters can be deceitfully low in a certain setting, yet danger is still quite present.

A good example would be an ultrafast laser with relatively weak energy per pulse. Imagine a 100-μJ laser with a 10-fs pulse width, along with 1 cm² spot size. The energy density here is quite low: less than 1 mJ/cm² in fact, which looks like it might be within the specifications of our baseline H absorber.

However, peak power density here would be 10 GW/cm², which is huge! In this example, a solution with W or VR absorber would have been much better (or the usage of QED attenuator with an energy meter). In this example, the relatively weak energy density level misled us into thinking we would be fine until we considered pulse width into the picture (and therefore peak power density).

Wall-plug Efficiency of Laser

The wall-plug efficiency of a laser system is its total electrical-to-optical power efficiency, i.e., the ratio of optical output power to consumed electrical input power. Taking the term seriously, the electrical power should be measured at the wall plug, so that this efficiency includes losses in the power supply and also the power required for a cooling system, which can be significant for high-power lasers. However, it is common that the wall-plug efficiency is calculated based on the electric power delivered to the laser diodes (e.g. in a diode-pumped solid-state laser system), ignoring losses in power supplies (which can be quite small for modern switched-mode power supplies).

Using the term in this common way, values of the order of 25% result for many diode-pumped laser systems ($\rightarrow$ all-solid-state lasers), e.g. Nd:YAG lasers. Even values above 30% are possible, e.g. with thin-disk lasers based on Yb:YAG and efficient laser diodes. It is to be expected that within the next few years laser diodes could become even more efficient, further raising the wall-plug efficiency of such systems. Pure laser diode systems can reach the highest efficiencies, sometimes well above 60%, but they cannot always be directly used, e.g. because of their poor beam quality and their inability to generate intense pulses. When using a high-power fiber laser as a brightness converter, one can obtain high output beam quality and (to some extent) intense light pulses, while the overall wall-plug efficiency can in the best cases be of the order of 50%. On the other hand, argon ion lasers, and even more so titanium–sapphire lasers and the like when they are pumped with argon ion lasers, generally have wall-plug efficiencies around or below 0.1%.

Particularly for high-power lasers, a high wall-plug efficiency is a very important quality. It reduces the electrical power consumption and also the amount of heat which has to be removed. Therefore, it not only cuts down the electricity bill but also reduces the demands on electrical installations and on the cooling system, and in turn often also the size of the laser system. Even for low-power lasers, the efficiency can be important in certain application areas, where the power budget is tight. Examples are telecom devices with a large number of transmitters, and lasers for space applications.

Slope Efficiency

An important property of an optically pumped laser is its slope efficiency (or differential efficiency), defined as the slope of the curve obtained by plotting the laser output versus the pump power. Usually, this curve is close to linear, so that the specification of the slope efficiency as a single number makes sense. However, nonlinear curves can occur under certain circumstances, e.g. as a consequence of quasi-three-level characteristics of the gain medium or thermal effects. For example, there can be a thermal roll-over, if the gain medium becomes hot at high pump powers, and this decreases the power conversion efficiency. A laser may even stop working for too high pump powers, for example when it leaves the stability zone of the laser resonator due to excessive thermal lensing. In case of such nonlinear curves, the slope efficiency is often determined from some approximately linear part.

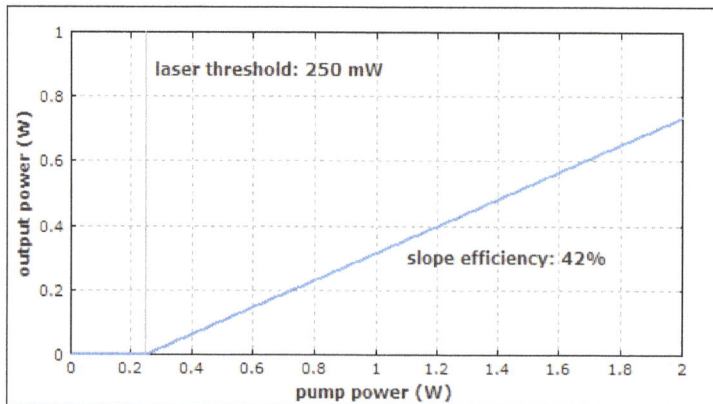

Laser action only occurs above a certain threshold pump power.
For higher pump powers, the output power often rises about linearly.
The slope of that line is called the slope efficiency.

The slope efficiency may be defined with respect to incident or absorbed pump power. For comparisons of power efficiency, it is usually fair to compare slope efficiencies with respect to incident powers, so that the pump absorption efficiency is taken into account. However, there are cases where values based on absorbed pump power are useful, e.g. for judging the intrinsic efficiency of the gain medium.

In simple situations (e.g., for some diode-pumped YAG lasers), the slope efficiency is essentially determined by the product of the pump absorption efficiency, the ratio of laser to pump photon energy (→ quantum defect), the quantum efficiency of the gain medium, and the output coupling efficiency of the laser resonator. For lamp-pumped lasers, it can be difficult to calculate the slope efficiency due to the difficulties of determining the fraction of pump power which is absorbed in the laser crystal, the position-dependent extraction efficiency and the complicated spectral dependence.

The optimization of the laser output power for a given pump power usually involves a trade-off between high slope efficiency and low threshold pump power. The optimum is usually a situation where the pump power is a few times the threshold pump power, and the slope efficiency is reduced below the value attainable with a stronger degree of output coupling.

The slope efficiency can also be defined for other laser-like devices such as Raman lasers and optical parametric oscillators. In the latter case, the differential slope efficiency with respect to incident pump power can even well exceed 100% under certain circumstances.

Quantum Efficiency

The quantum efficiency (or quantum yield) is often of interest for processes which convert light in some way. It is defined as the percentage of the input photons which contribute to the desired effect. Examples are:

- In a laser gain medium, the pump process may require the transfer of laser-active ions from one electronic level (into which the ions are pumped) to the upper level of the laser transition. This pump quantum efficiency is the fraction of the absorbed pump photons which contributes to the population of the upper laser level. This efficiency is close to unity (100%) for many laser gain media, but can be substantially smaller for others. It may depend on factors like the excitation density and parasitic absorption processes. It is not easy to measure, since the power conversion efficiency also depends on other factors, such as optical losses.

- Similarly, the quantum efficiency of fluorescence can be defined. It can be reduced by non-radiative processes such as multi-phonon transitions and energy transfer processes. If such effects do not occur, it can be essentially 100%.

- In a photodiode (or some other photodetector), the quantum efficiency can be defined as the fraction of incident (or alternatively, of absorbed) photons which contribute to the external photocurrent. In the visible and near-infrared region, photodiodes can have quantum efficiencies above 90%, although values between 40% and 80% are more common. Photomultipliers can have much lower quantum efficiencies, strongly depending on the wavelength region. In case of avalanche photodiodes in Geiger mode, dead time effects are not considered for the quantum efficiency.

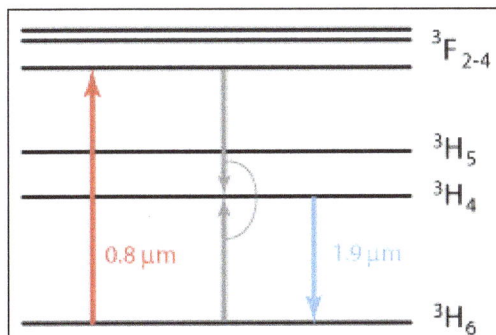

1.9-μm emission in a thulium-doped fiber laser with > 100% quantum efficiency.

In some special cases, the quantum efficiency of a laser or laser amplifier can be larger than unity. This is due to certain energy transfer processes between laser-active ions, which lead to a kind of cross-relaxation: starting with one ion in some excited state, a part of its energy is transferred to some other ion, which was originally in the electronic ground state, and both ions are finally in the upper laser level. This can, of course, only happen when the photon energy of the laser transition is lower than half that of the pump light. An example, illustrated in figure, is that of thulium-doped 1.9-μm fiber lasers, where ions are pumped into the level $^3F_{2-4}$, and a cross-relaxation process (gray arrows) populates the upper laser level $^3H_4$. This could in principle lead to a quantum efficiency of up to 200%. Values well above 100% can be reached in practice.

Methods for Laser Spectral Characterization

Many laser applications, including high-resolution spectroscopy, optical remote sensing, cooling/trapping, and optical fiber communications, are highly dependent on a laser's spectral performance. Therefore, it is important to characterize laser parameters such as absolute wavelength, wavelength stability, linewidth and longitudinal mode structure. There are a variety of instruments that can provide this information, but most systems do not provide a complete picture. Wavelength meters measure absolute wavelength very accurately, but provide little or no spectral information. To determine a laser's spectrum (power vs wavelength), a spectrum analyzer is necessary. These systems have high resolution, but their wavelength measurement accuracy is limited.

Wavelength Meters

Wavelength meters use optical interferometry to measure a laser's absolute wavelength very accurately. Measurements are made in real-time resulting in the ability to automatically report and control laser wavelength. Two types of interferometers are used. These are the Michelson interferometer and the Fizeau etalon.

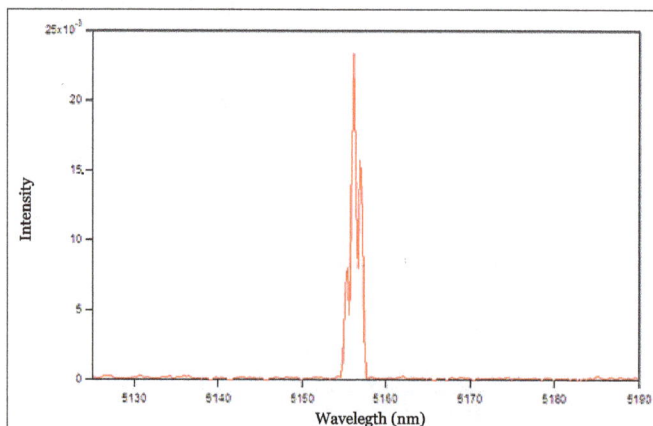

The Michelson interferometer with FFT analysis is very flexible with the types of lasers that can be analyzed, especially with regard to the laser's wavelength. Since the Michelson interferometer uses a single element photodetector, a variety of such photodetectors are readily available for operation from the ultraviolet into the mid-infrared.

Michelson Interferometer

The scanning-mirror Michelson interferometer generates information from the interference of two beams that originate from the same source. The optical input is split between a fixed path and a path that is smoothly changing in length. Both beams are reflected and recombined to produce an interference pattern that is a consequence of the changing phase relationship between the beams. This temporal interference pattern is analyzed by counting the number of interference fringes generated during the scan of the mirror. This results in the ability to measure absolute laser wavelength to an accuracy as high as ± 0.0001 nm.

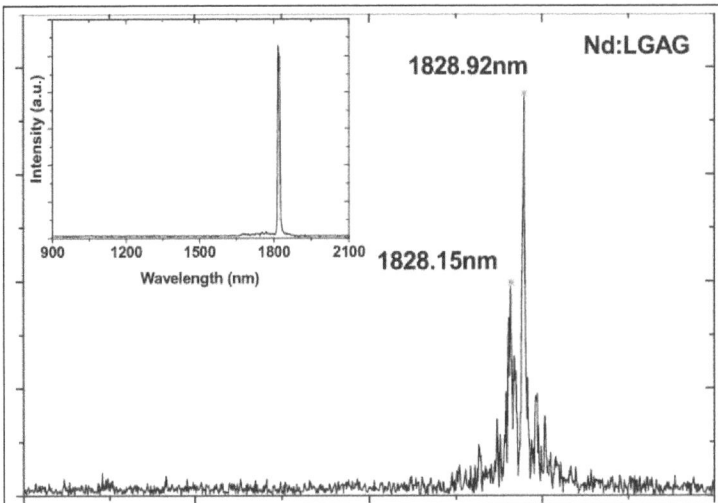

It shows the spectrum of a quantum cascade
laser (QCL) that operates at about 5.1 m.

Fizeau Etalon

This type of wavelength meter uses fixed-spaced Fizeau etalons to generate a spatial interference pattern. Data is collected by imaging the interference patterns onto a photodetector array. The spacing between the interference fringes is measured and then used to calculate absolute laser wavelength. In general, the Fizeau etalon-based wavelength meter is used to provide wavelength information only.

Figure shows a "simulated" DWDM signal that is generated by sending a broad ASE source through a Fabry-Perot etalon with a free spectral range of 100 GHz. The entire spectrum of the etalon transmission is shown, giving the spectral profile of the ASE source over the wavelength range of about 1450 to 1650 nm.

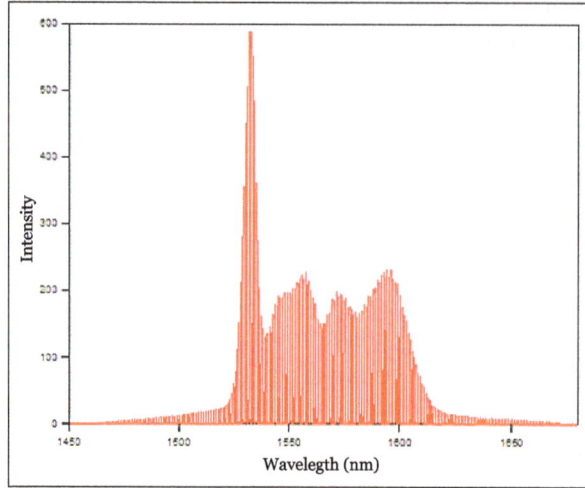

Spectrum Analzyers

A spectrum analyzer is used to determine a laser's spectrum, which is a measure of laser power vs wavelength. Two techniques are commonly used in commercially available systems. These are the Fabry-Perot interferometer and the diffraction grating.

Fabry-perot Interferometer

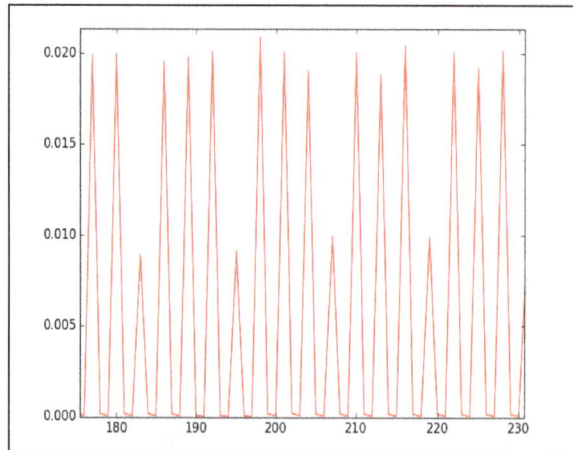

Figure shows a magnified portion of this spectrum. The "simulated" optical channels (individual peaks) are shown with a frequency separation of 100 GHz.

This is a simple device that relies on the interference of multiple beams. It consists of two partially transmitting mirrors precisely aligned to form a reflective cavity. Incident light enters the Fabry-Perot interferometer and undergoes multiple reflections between the mirrors so that the light interferes with itself many times. This results in an interference pattern that is characteristic of the laser's power vs wavelength spectrum. The primary advantage of this type of laser spectrum analyzer is its spectral resolution which can be better than 10 MHz, or 0.00003 nm at 1000 nm. Therefore, it is ideal for measuring laser linewidth, longitudinal mode structure, and wavelength stability. However, the Fabry-Perot interferometer only provides relative information so it will not measure absolute laser wavelength.

Optical Spectrum Analyzer (OSA)

Instruments identified as optical spectrum analyzers, or OSAs, typically employ diffraction grating technology. The diffraction grating is a dispersive element that splits the incident light into several beams travelling in different directions. The number and direction of these beams depend on the wavelength of the incident light. The diffracted beams are imaged onto a photodetector array for analysis. The primary advantage of such a spectrum analyzer is its ability to make precise measurements on the power axis of the spectrum. An OSA is, in effect, a wavelength-selective optical power meter. That is, it can measure the absolute power of the spectral components very accurately. It also has excellent sensitivity and a low noise floor resulting in the ability to measure low power optical signals. However, the OSA's ability to measure absolute wavelength is limited to an accuracy of about ± 0.01 nm, and a typical spectral resolution is only about 0.07 nm.

References

- Laser-esonators: rp-photonics.com, Retrieved 13 February, 2019

- Laser-modes, laser-tutorial, studies: fraunhofer.de, Retrieved 17 March, 2019

- Longitudinal-modes, lasers: scitec.uk.com, Retrieved 23 June, 2019

- Linewidth: rp-photonics.com, Retrieved 12 April, 2019

- Slope-efficiency: rp-photonics.com, Retrieved 21 March , 2019

- Complete-laser-spectral-characterization: photonics.com, Retrieved 13 June, 2019

- Beam-quality: rp-photonics.com, Retrieved 24 March , 2019

4

Applications of Lasers

Lasers are applied in numerous fields for a variety of purposes. Some of these are satellite laser ranging, communication, laser cutting, laser welding, scanning for building design and construction, and printing. These diverse applications of lasers have been thoroughly discussed in this chapter.

Laser Applications in Defense

Laser Range Finder

To knock down an enemy tank, it is necessary to range it very accurately. Because of its high intensity and very low divergence even after travelling quite a few kilometres, laser is ideally suited for this purpose. The laser range finders using neodymium and carbon dioxide lasers have become a standard item for artillery and tanks. These laser range finders are light weight and have higher reliability and superior range accuracy as compared to the conventional range finders.

The laser range finder works on the principle of a radar. It makes use of the characteristic properties of the laser beam, namely, monochromaticity, high intensity, coherency, and directionality. A collimated pulse of the laser beam is directed towards a target and the reflected 1ight from the target is received by an optical system and detected. The time taken by the laser beam for the to and fro travel from the transmitter to the target is measured. When half of the time thus recorded is multiplied by the velocity of light, the product gives the range, i.e., the distance of the target.

The laser range finder is superior to microwave radar as the former provides better collimation or directivity which makes high angular resolution possible. Also, it has the advantage of greater radiant brightness and the fact that this brightness is highly directional even after travelling long distances, the size of the emitting system is greatly reduced. The high monochromaticity permits the use of optical band pass filter in the receiver circuit to discriminate between the signal and the stray light noise.

A typical laser range finder can be functionally divided into four parts: (i) transmitter, (ii) receiver, (iii) display and readout, and (iv) sighting telescope. An earlier version of a laser range finder is schematically shown in figure. The transmitter uses a Q-switched Nd:YAG laser which sends out single, collimated and short pulse of laser radiation to the target. A scattering wire grid directs a small sample of light from the transmitter pulse on to the photodetector, which after amplification is fed to the counter. This sample of light starts the counter. The reflected pulse, received by the telescope, is passed through an interference filter to eliminate any extraneous radiation. It is then focused on to another photodetector. The resulting signal is then fed to the counter. A digital system converts the time interval into distance. The range, thus determined by the counter, is displayed in the readout. The lighting telescope permits the operator to read the range while looking at the target.

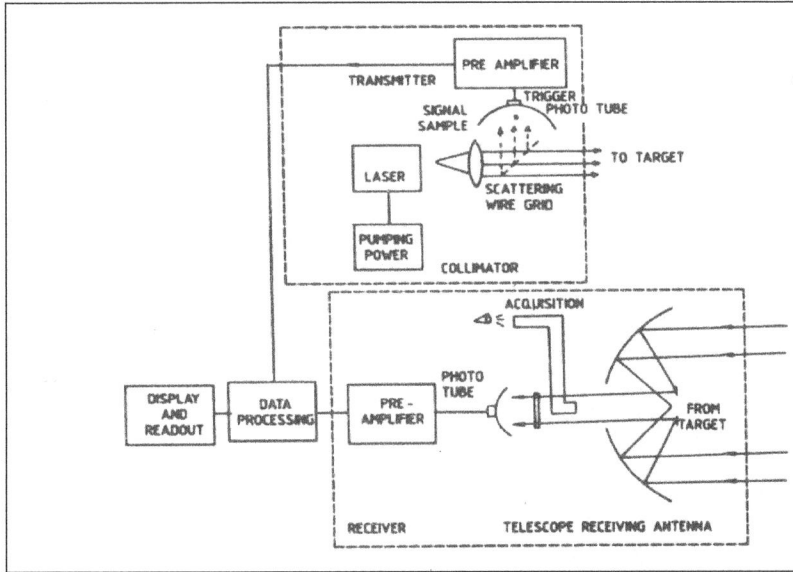

Special circuits have been used to eliminate Spurious signals with the help of range gating and to make the use of laser range finder. Possible under all weather conditions for which the targets can be seen visually through the sighting telescope. The modern versions of the laser range finders use either high repetition pulsed Nd:YAG laser or carbon dioxide laser with range gating system. In ranging a target about 10 km away using these systems, an accuracy within 5 m is easily obtained. The laser range finders of medium range (up to 10 km) are used in several Defence areas, including:

- Tank laser range finder for artillery, an armoured vehicle, or a truck.

- Portable laser range finders, used in the field artillery fire control systems. These are intended for field application in conjunction with artillery fire control systems.

- Airborne laser range finder, pod-mounted and servo-positioned for the Air Force. In any airborne weapon system, one of the i.e., the distance of the target. The laser range finder combines the characteristic features of a laser with gyroscope stabilisation to provide an equipment which is more accurate and has a faster response than any other means of deriving air-to-surface or air- to-air range. At the same time, it is more compact than any radar.

- The laser walkie-talkie range finder, a compact small instrument, weighing less than 4 kg, useful to range objects at distances less than 5 km. This range finder uses the semiconductor diode laser in emitting short duration pulses. With this, it is possible to which transmit and receive audio/visual communications, or pinpoint targets with a hand-held laser, even from unsteady environment in a helicopter or on a ship being tossed around by the rolling seas. There are no separate tripods, unwieldly power packs, or other external accessories. It gives an immediate readout of distance and elevation right on the instrument.

Underwater Laser

Lasers can also be used as a source of underwater transmission. For this purpose, a laser giving radiation in the blue-green region is most suitable as the transmission in this region is maximum for sea water. The attenuation in underwater transmission is due to (i) absorption by materials in water, (ii) scattering by suspended particles, and (iii) variation in optical density along the light path. The blue-green lasers have assumed much importance in the systems related to naval applications.

At present, the submarines have to rely on a sonar to find the enemy crafts and to avoid the underwater objects. This has serious limitations. The whales, dolphins and other marine life give false signals. A typical sonar cannot give a well-defined picture because the sonar beam is broadened or scattered by sea water. A difference in the saltiness of water can cause the sonar beam to bend and make the target appear where it is not. Another problem of using sonar is that it gives away to the enemy the position of the ship from which it is transmitted.

Lasers can be used efficiently for ranging and detection of underwater objects. For this purpose, a frequency doubled Nd:YAG laser or an argon ion gas laser or a Raman shifted xenon chloride laser is used. A schematic diagram of an underwater ranging and viewing system is shown in figure. It consists of the laser transmitter which sends high power laser pulses of about 10 ns duration to the target at the rate of 30 to 50 per second through a beam splitter and a diffuser small amount of the laser light reflected by the beam splitter is made to fall on the photodiode the ranging and display circuit to start the time interval counter. The reflected light from the target is collected by telescopic optics after stray radiation is eliminated by an interference filter. A range gating circuit helps to avoid unwanted echoes. The reflected pulse from the target is intensified by the image intensifier and the output is fed to image orthicon, which gives the display of the object. In this way, both the range and the image of the target are obtained. With high power release of several megawatts power, underwater ranging is possible up to 500 m in clear water.

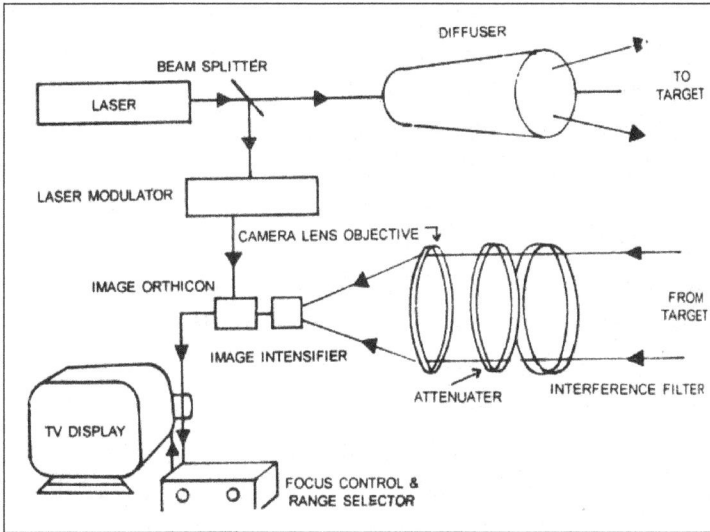

Schematic diagram of underwater ranging.

Lasers can also be used for communication between submarines ensuring absolute privacy and in guidance systems for torpedoes and other unmanned underwater vehicles. Recent underwater laser communication has been established via satellite, i.e., from ground-to-satellite and then to underwater station.

Laser-guided Anti-tank Missile (ATM)

A missile can be guided and controlled by an infrared beam emitted from a laser, with extremely small divergence. This can be achieved in four ways:

1. The laser beam is used to illuminate the target tank; the anti-tank missile (ATM) then homes on to the target, as the latter has become a source of back-scattered radiation.

2. The laser beam is used to provide guidance instructions to the missile, i.e., it provides the command link.

3. The missile rides the laser beam which is kept pointing along the collision course to the target.

4. The missile itself carries a laser scanner and seeker for active homing on to the target.

In the first case, the laser target designator is a pulsed Nd:YAG laser. The laser beam is so modulated that the receiver, a four quadrant detector in the missile, is able to calculate any divergence of the missile trajectory from the beam axis and correct the deviation by altering the fins of the missile. The guidance unit consists of both optical and electronic equipment. This enables the gunner to aim the infrared guidance beam for firing the missiles.

The system in which the missile is a beam rider designed to ride the laser pointing in the direction of the target, is more attractive. Missile can carry four detectors at the wing tips looking towards the rear of the missile. The detectors determine the central axis of the laser beam and keep the flight path of the missile along it. The wavelength of the laser should be such that is the least absorbed by the plume of the sustainer motor. Thus in a laser designator, the laser by virtue of its narrow beam illuminates a chosen target. A receiver in a bomb or a missile seeks the target illuminated from the scattered laser radiation and homes on to it. In the Vietnam War and in the recent Iraqi war, the Americans used laser guided missiles with pinpoint accuracy to destroy the enemy targets.

Ring Laser Gyroscope

The ring laser gyroscope is an extremely useful instrument for sensing and measuring very small angles of rotation of the moving objects. It has now replaced the mechanical gyroscopes used in most of the aircraft (both civil and military) he and also in long range guided missiles. The main advantages of the ring laser gyroscope are: (i) non-existence of moving parts, (ii) high g capability, and (iii) higher reliability as compared to the mechanical gyroscope. In addition, the laser gyroscope is capable of wide dynamic range and rapid reaction time, the characteristics required for missile guidance.

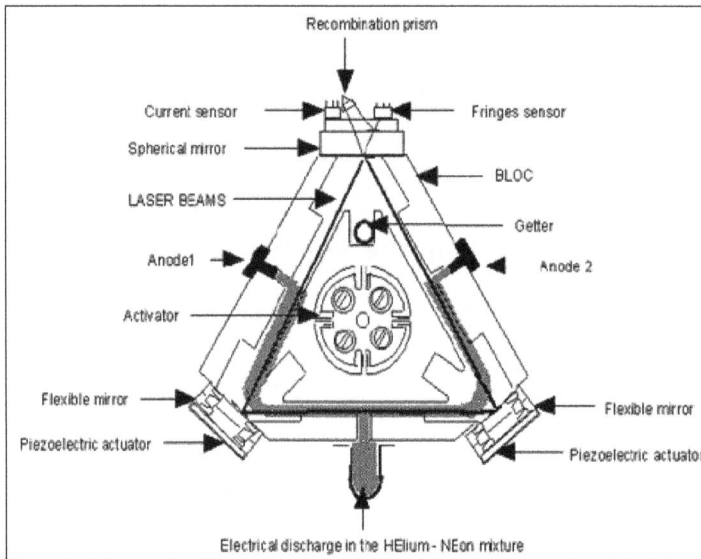

Recombination prism
Current sensor — Fringes sensor
Spherical mirror
BLOC
LASER BEAMS
Getter
Anode1 — Anode 2
Activator
Flexible mirror — Flexible mirror
Piezoelectric actuator — Piezoelectric actuator
Electrical discharge in the HElium- NEon mixture

The ring laser gyroscope basically consists of a ring cavity around which two laser light beams travel in opposite directions. The operation of the ring laser gyroscopes is based on the so called Sagnac effect by which rotation of an object is sensed by an interferometric technique.

In a triangular cavity of a quartz block, laser beam is split into two light beams with the help of suitable mirrors. These two light beam travel in opposite directions in the same path of the cavity, one in the clockwise and the other in the anti-clockwise direction.

The two light beams then pass through a beam splitter and a beam combiner, behind which a readout detector is placed. If the cavity which is acting as an interferometer is stationary, the two light beams travel the same distance in the opposite directions without any path difference and hence no interference takes place. However, if the block is rotated clockwise about an axis through the centre and perpendicular to the plane of the interferometer, the beam travelling in the clockwise direction travels a path length slightly more than the beam travelling in the anti-clockwise direction. When these two light beams recombine at the beam combining prism, interference takes place due to the path difference; the interference fringes displaced in the field of view are proportional to the amount of rotation of the block. The laser gyroscope uses a helium-neon gas laser to generate monochromatic radiation in the two directions inside the triangular quartz block. Two photodetectors sense the direction and the rate of rotation. The output is proportional to the input angle. The whole system is a single plane, rate integrating gyroscope and is capable of measuring rotation rates of the order of 1/10,000 degree/hour.

The main use of the ring laser gyroscope is for inertial navigation. It is being used in inertial guidance of aircraft, ships, and missiles; flight control; and gun-fire pointing. Both Honeywell and a Litton Industries, USA, the manufacturers of the ring laser gyroscopes, have introduced them in the Boeing 757 and 767 and Airbus A 310, now in production. These gyroscopes are ideally suitable for the various guided missile applications the Defense sector also.

Air Reconnaissance

Lasers can be used as a secretive illuminators for aerial reconnaissance during night with high precision. Earlier it was done using a camera, equipped with either magnetic flares or powerful strobe lights with their cumbersome power supplies. For this purpose, a helium-neon laser or a gallium arsenide semiconductor laser is used. Two properties of the laser, namely, its narrow beam and its radiance or brilliance are of importance in this particular application.

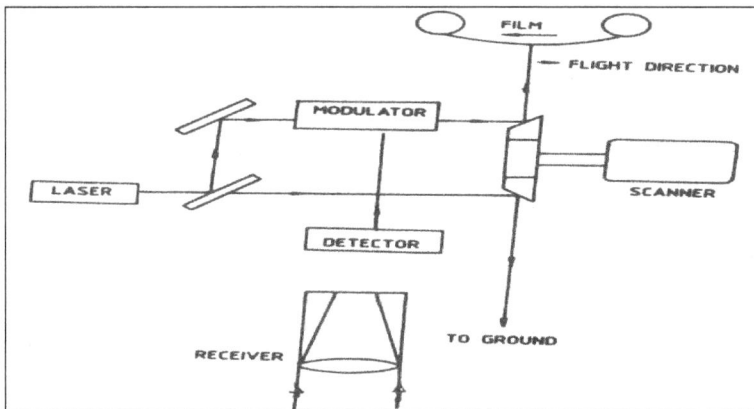

The block diagram of the laser camera is shown in figure. One of the beams passes downwards through a six-sided prism scanner towards the earth. The prism scans through a selected angle at right angles to the direction of the flight of the aircraft. The other beam passes through a Pockels cell modulator. On emerging from the modulator, the beam strikes the prism scanner and is then reflected towards and recorded on the film.

The laser beams reflected from the target area are picked up by a Schmidt lens, which images the light on to a photodetector. The video output of the photodetector, corresponding to the reflectivity of the observed terrain, drives the modulator. Thus, the returned beam modulates the original beam. The pictures thus obtained are comparable in resolution with those taken under daylight conditions. Thus, the enemy targets can be photographed at night under high secrecy during the flight of the aircraft. The laser camera system was tested successfully by the United States Air Force Tactical Air Reconnaissance Centre.

Anti-missile Defence System (Star Wars)

In an antimissile defence system, laser is used to dispose the energy of warhead, not by vaporising or melting it, but by partially damaging the missile, say by drilling a hole. Tremendous energy is required to completely burn the missile, which is not practicable. If a guided vane of a missile is fractured, several vibrations will be developed in the air frame thereby disintegrating major sensitive portion of the missile.

Two types of anti-missile defence systems have been visualised. One such system, laser kill system is completely earthbound. Here, an early warning microwave radar gives a rough position of the approaching missile. Then a lidar aligned to the target by the tracking radar gives the precise position of the missile. This data is fed on to another high intensity laser beam which actually does the killing. To exploit the laser's killing capability, a high speed servo system and a complex focusing system are essential.

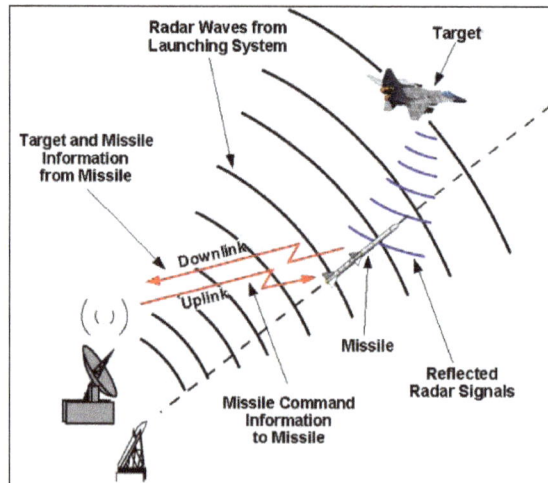

The other anti-missile defence system is the orbiting space station, equipped with detecting, tracking and killing laser devices. An infrared homing system on the laser weapon is used to close on an enemy vehicle and then fire a high energy laser beam.

Firing by laser weapons would not change the positional or altitude stability of the space station. It is predicted that the lasers would ultimately make inter-continental ballistic missiles (ICBM's) obsolete.

There are, however, many limitations in the utilisation of laser in its anti-missile role. The power required is very prohibitive and as a result huge power stations are required for the operation. At present, huge power of more than 100 kW in the continuous mode is being obtained from the gas dynamic carbon dioxide lasers and some other chemical lasers, developed in the US and Russia. This amount of laser power is sufficient to destroy an enemy vehicle or a missile.

The SDI or Star Wars is the US programme aimed at defending itself and its allies against the ICBM strikes through space. The concept of strategic defence is concentrated at a three-layered defence system in which the enemy missile can be destroyed in the boost, or the mid-course, or terminal phase. Each layer of defence employs several alternatives of weapon systems. Laser is one such important system.

For detecting and destroying missiles, different types of high power lasers, such as gas dynamic carbon dioxide, excimer, x-ray, free-electron, and chemical lasers can be used. In one such programme, the us scientists are developing 5 to 10 MW deuterium fluoride lasers for destroying the ICBM. In the ground-based laser systems, laser beams will be directed towards a large mirror in geosynchronous orbit. From there, the beams will be directed towards a moving mirror in low orbit which will reflect the beams on to the missile to destroy it. The nuclear-pumped x-ray lasers are also being considered for destroying missiles in the boost phase.

Laser Proximity Fuze

The proximity fuze, developed in the US using a solid-state laser, detonates the missile warhead when it comes within the range of its target. The higher manoeuvrability of the missile is expected to improve its performance a great deal in close in aerial combat. It is also claimed that the proximity fuze and the warhead will enable the missile to destroy its target without hitting it directly.

Laser Beacon

The present infrared light sources, being used as ground beacons to identify the ground points, are inefficient and not much reliable. Using a lensless diode array, a laser beacon can made multi-directional. The laser beacons are light in weight, efficient and have long life. Another advantage is that the pulses can be used ground beacons so that the airdropping of supplies can be done at the given locations efficiently.

Weapon Firing Simulator

If we take into account the cost of ammunition and large land that is required to fire it, basic training of the tank gunners is very expensive. A simulator technique, which does not sacrifice the acquisition of the basic skills during trials, has been developed in which the main weapon has been replaced by a laser. The technique is known as the weapon firing simulator.

The simulator, installed in the machine gun mount of the vehicle, produces a single burst intense red light when the firing circuit of main weapon is activated. On seeing this bright spot of light, which is visible momentarily in optical system of the vehicle, the trainee is able to lay his gun sights accurately to track the targets. It also enables one to check the accuracy and proficiency of the crew in operating the vehicle's main weapons system.

Lidar

When the laser beam is used for a radar application, it is called lidar. The details, which could not be achieved earlier with microwave radars, can now be obtained with lidar. Besides, the laser beam can be focused with lenses a mirrors easily whereas microwaves need huge antenna for focusing. As a beacon or a radar, the advantages of utilising small antenna and components are obvious. With a lidar, the dimension and the distance of the target can be obtained with higher accuracy, which is not possible with the conventional microwave radar. The lasers used in lidars are of carbon dioxide, Q-switched neodymium, or gallium arsenide semiconductor type.

The great advantage of the use of carbon dioxide lasers for radar application is their capacity to produce high power output with requisite. The spectral purity. The coherent carbon dioxide laser tips radar functions essential like a coherent microwave radar except for the fact that the carbon dioxide laser beam has a frequency of a few thousand times more than that of the X-band radar and at it a sharp beam width of a few microradians. The high frequency of the carbon dioxide laser also produces high Doppler shift even from slow-moving targets. The fine beam width and high Doppler the shift give the carbon dioxide laser an unparalleled imaging capability. This radar system is used for and measuring radial velocities to track low-flying aircraft and slow-moving objects. Since the laser beam is very much attenuated by rain, fog, or snow, the lidar can perform well only in good weather conditions. When it comes to lasers for lidar, one size doesn't fit all: system designers need to understand the physical environment and performance objectives of their application before source selection can be attempted.

Merger and acquisition activity is rampant for light detection and ranging (lidar) system and component manufacturers serving the autonomous vehicle industry, but other industries are joining the lidar frenzy.

"The agriculture industry is using lidar to map terrains and to guide farmers on the selective distribution of fertilizers and pesticides. In biology and conservation, lidar monitors forest canopy heights or deforestation. The military uses lidar for autonomous vehicle and drone guidance, for identifying possible targets, and for tactical precision mapping. And, in the mining industry, lidars map excavated areas and determine volume removal. "For surveying, lidar maps buildings and surrounding areas for future development, and atmospheric lidar systems map the 3D distribution of aerosols and molecules to characterize everything from pollution to clouds, wind, and even volcanic emissions."

An Aerosol Multiwavelength Polarization Lidar Experiment (AMPLE) multi-wavelength lidar system uses 355, 532, and 1064 nm lasers with three individually addressable output channels with 4 mJ energy, 1 kHz repetition rate, and 1 ns pulse duration to monitor volcanic emissions from Mount Etna: (a) The temporal evolution of aerosol-induced depolarization (b) measured by AMPLE is used to monitor mineral dust in the first 6 km above ground.

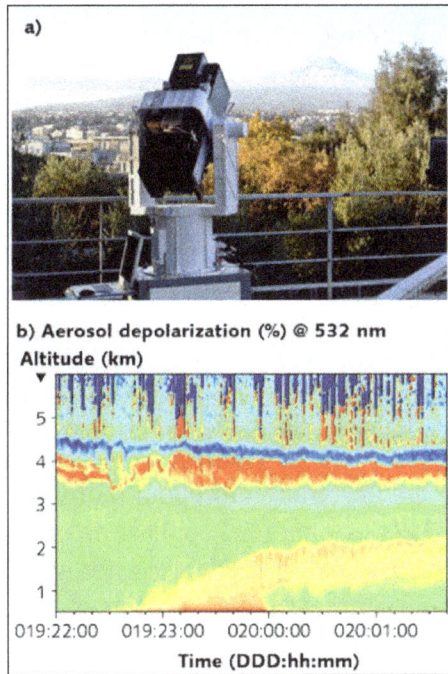

a)

b) Aerosol depolarization (%) @ 532 nm

Altitude (km)

Time (DDD:hh:mm)

Lidar systems illuminate a target area or scene with pulsed laser light and measure how long it takes for reflected signals to be returned to a receiver. A lidar system includes a laser source or transmitter, a sensitive photodetector or receiver, synchronization and data processing electronics, and either motion-control equipment or solid-state microelectromechanical systems (MEMS)-based components for precise laser scanning to create 3D maps and/or proximity data.

Among these required components, Smith reminds us that the laser itself contributes to overall system performance. "The laser beam quality and divergence, for example, is responsible for the lateral resolution [x, y] of mapping lidars, while short pulse duration and timing jitter are responsible for the longitudinal accuracy [z]." Pulse energy is the key parameter to attain long ranges, while high pulse repetition rates allow faster scanning and high data throughput.

Performance Considerations

"High-peak-power (tens of kilowatts to tens of megawatts) pulsed (nanosecond realm) solid-state lasers have been used in lidar applications for decades." "Size and weight, cost, power consumption, liquid cooling, sensitivity to shock and vibrations, and harsh environments have limited the diffusion of lidar instruments in mobile, airborne, and space applications. But companies like Bright Solutions [Pavia, Italy] have recently developed a new generation of high-peak-power, sub-nanosecond, air or conduction-cooled, Q-switched solid-state lasers that remove these limitations and offer a wide range of laser wavelengths from the ultraviolet (UV) to the near-infrared (near-IR)."

For airborne topographic mapping, a wavelength around 1 μm is typically used, in which case the beam is expanded large enough to be considered eye-safe. For bathymetry (high-resolution mapping of the sea bottom and coastal areas), a high-energy, frequency-doubled 532 nm laser sources is often used, as the green wavelength represents the best compromise between high transmission in pure water and limited backscattering from submarine particulates.

The 532 nm to 1 μm wavelengths have benefits in terms of cost and energy consumption, but the need to reach relatively long ranges can easily lead to increased laser power over the limit of Class 1 laser safety. In such a case, laser emission can be hazardous to the eye if the beam is not expanded to an acceptable diameter (thus increasing the system size) to make it comply with eye-safe regulations.

Intrinsically eye-safe lasers are becoming more and more popular in high-performance compact lidar meant for civil and commercial applications. Among eye-safe wavelengths, IR lasers emitting around 1.5 μm are often the choice when solid bodies need to be detected as in topography mapping and obstacle avoidance. In fact, the atmosphere is very transparent and detectors are very efficient at 1.5 μm. Alternatively, UV wavelengths around 355 nm or shorter are the best choice for eye-safe atmospheric lidar systems, where relatively high backscattering coefficients from the atmospheric particulate are desired.

Beyond wavelength considerations, what about pulse duration? "Ideally," "lidar designers would like to reach millimeter-to-centimeter longitudinal measurement resolution, so short pulse durations should be considered. Nevertheless, very short pulses—on the order of a few picoseconds—would cause the laser spectrum and the receiver bandwidth to broaden, thus worsening signal-to-noise ratio." Pulses longer than 1 ns would result in lower noise but poorer resolution, making pulse durations on the order of a few-hundred picoseconds (or sub-nanosecond) the best compromise for high longitudinal accuracy and signal-to-noise ratio.

To demonstrate that one size doesn't fit all when it comes to lasers for lidar, let's explore two very different applications—autonomous vehicles and forest canopy mapping.

Autonomous Vehicle Lidar

According to a recent survey from ABI Research, the number of devices sold for automotive vehicles should reach 69 million units in 2026. As explained by Frédéric Chiquet, R&D manager Guillaume Canat, and CEO Marcle Flohic of Keopsys Group, there are two main types of autonomous vehicle lidar systems: 3D flash lidar and scanning lidar.

Flash lidar uses a wide-angle-emitting source and wide-angle optics (a fisheye lens, for example) to focus backscattered light acquired during a single emission onto a matrix detector to obtain all the time-of-flight (TOF) data needed to model the area surrounding a vehicle. Conversely, scanning lidar addresses the 3D environment line by line; light is sequentially emitted in each direction and the corresponding echoes are detected one by one by the detector. The eye-safe laser source must operate in pulsed mode, be powerful enough to detect a pedestrian wearing dark clothes at 100 m, operate from -40 to 85 °C, and emit pulses compatible with a measurement distance accuracy of 10 cm.

While many lidar sources are laser-diode-based, uncooled fiber lasers are also used and offer many advantages over pulsed laser diodes, including the ability to have their high-power beams split and routed to multiple sensor locations using optical fiber. Using a master oscillator power amplifier (MOPA) architecture, a typical 1550 nm lidar fiber laser has pulse repetition rates of 5 to 250 kHz at power levels from 10 to 15 kW and 200 to 300 W, respectively.

Pulsed laser diode and fiber laser sources are compared for autonomous vehicle lidar applications.

Features	Pulsed laser diode	Fiber laser
Peak power	+	+++
Pulse repetition frequency	+	+++
Pulse duration	++	+++
Beam quality	+	+++
Beam divergence	+	+++
Compactness	+++	+
Reliability	+++	+++
Multi-emission pooling	+	+++
Optical arrangement ease	+	+++
Eye safety	+	++

The pulsed laser diodes dedicated to autonomous cars are hybrid devices. A laser chip is mounted with capacitors that are triggered by a MOSFET transistor. Each time that the transistor gate is opened, the electric charge accumulated in the capacitors is discharged into the chip, which emits the optical pulse. But although this type of source is cost-effective, as its 905 nm output is easily detected by silicon detectors rather than expensive 1550 nm. In GaAs photodiodes, laser diodes have limited pulse repetition rate and lower peak power, and are limited by overheating effects.

Laser diode sources for 3D flash lidar are based on diode stack technology, with several edge-emitting bars assembled into a vertical stack, with each layer separated by a thin heat sink to limit internal overheating. Unfortunately, non-coherent addition of the stacks creates high output power that often fails to meet Class 1 eye-safe requirements and while vertical-cavity surface-emitting lasers represent a more cost-effective alternative to stacks, their weaker output power limits their use to shorter-range TOF applications.

Forest Canopy Mapping Lidar

Researchers Supriya Chakrabarti, Timothy Cook, Kuravi Hewawasam, and Glenn Howe in the Lowell Center for Space Science and Technology (LoCSST) at the University of Massachusetts, Lowell, as well as colleagues from the University of Massachusetts, Boston and Boston University, have developed and field-tested a new terrestrial laser scanning lidar instrument built for the purpose of recording and retrieving the 3D structure of forest vegetation.

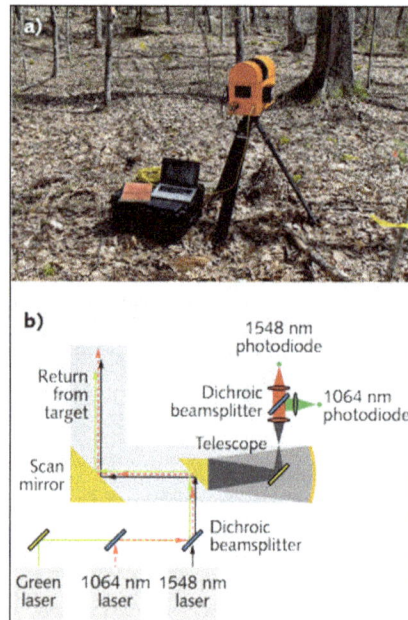

The DWEL lidar instrument in operation at the Harvard Forest in Petersham, MA
(a) Includes all optics and readout electronics in its orange box
(b) As well as a laptop computer to manage data from the field.

This Dual-Wavelength Echidna Lidar (DWEL) uses the differential reflectances of vegetation targets as observed from simultaneous laser pulses—one in the shortwave-infrared

(SWIR) range and the others in the near-IR region—to characterize forest structure by distinguishing between leaves and trunks at an angular resolution as fine as 1 mrad.

Based on Echidna laser scanning technology patented by Australia's Commonwealth Scientific and Industrial Research Organization (CSIRO), DWEL uses a rapidly rotating zenithal scan mirror and a slowly rotating azimuthal platform to provide full coverage of the angular scan space, with a 6 mm collimated beam diverging at 1.25, 2.5, or 5 mrad. The resulting field of view spans 360° in azimuth angle and ±117° in elevation angle. This full-waveform (as distinct from first-return) lidar completes a full-hemisphere scan by rotating 180° in azimuth in approximately 36 minutes in its standard resolution mode (2 mrad).

Including a green laser that allows the operator to visualize the beam direction, DWEL quantifies vegetation structure for environmental applications using the principle that leaves and other water-containing materials found in the forest absorb more strongly in SWIR (1548 nm) than near-IR (1064 nm), allowing the return signals of leaves to be separated from those of trunks, branches, and other dry materials in the forest.

The scene in the 1548 nm channel of the DWEL instrument (a) and the same image obtained in the 1064 nm channel (b) are shown.

Nominal laser pulse length, 5 ns full width at half maximum (FWHM), is specified to provide a full description of return pulse shapes as digitized at 2 GHz. Full-waveform recording allows detection of multiple targets out to 100 m from the scanner. Pulse energy of 0.6 µJ provides a good signal-to-noise ratio while maintaining eye safety criteria. A dark Lambertian target inside the instrument housing is used for calibration with each scan mirror rotation.

Return signals are collected by a 10-cm-diameter Newtonian telescope, where the beams are chromatically separated and out-of-band wavelengths are eliminated using narrow bandpass interference filters. Return signals from each laser are detected by a pair of detector-amplifiers using InGaAs photodiodes.

The power of DWEL over most other lidar systems is its two-wavelength vegetation discrimination capability: leaves, which contain more moisture than woody material, appear darker in the 1548 channel compared to the 1064 channel. For this particular lidar task, as in all others regardless of industry, the laser itself is critical to satisfying the unique performance requirements of the application at hand.

Satellite Laser Ranging

In Satellite Laser Ranging (SLR) a global network of stations measure the instantaneous round trip time of flight of ultrashort pulses of light to satellites equipped with special reflectors. This provides instantaneous range measurements of millimeter level precision which can be accumulated to provide accurate orbits and a host of important science products.

- Satellite Laser Ranging is a proven geodetic technique with significant potential for important contributions to scientific studies of the Earth/Atmosphere/Oceans system.

- SLR is the most accurate technique currently available to determine the geocentric position of an Earth satellite, allowing for the precise calibration of radar altimeters and separation of long-term instrumentation drift from secular changes in ocean topography.

- SLR's ability to measure the temporal variations in the Earth's gravity field and to monitor motion of the station network with respect to the geocenter, together with the capability to monitor vertical motion in an absolute system, makes it unique for modeling and evaluating long-term climate change by:

  o Providing a reference system for post-glacial rebound, sea level and ice volume change.

  o Determining the temporal mass redistribution of the solid Earth, ocean, and atmosphere system.

  o Monitoring the response of the atmosphere to seasonal variations in solar heating.

- SLR provides a unique capability for verification of the predictions of the Theory of General Relativity.

- SLR stations form an important part of the international network of space geodetic observatories, which include VLBI, GPS, DORIS and PRARE systems.

- On several critical missions, SLR has provided failsafe redundancy when other radiometric tracking systems have failed.

- The cost effectiveness of SLR operations are improving through increased standardization, configuration control, and automation. NASA is vigorously pursuing the development of SLR 2000, a stand-alone, low-cost, subcentimeter SLR system.

- The International Laser Ranging Service has been formed by the global SLR community to enhance geophysical and geodetic research activities.

The basic principles of SLR are quite simple. Figure describes the main components of an SLR system. A satellite equipped with a corner cube reflector (or an array of CCRs) is tracked by an optical telescope which has a sensitive light detector at its receiving end. In parallel and co-aligned, a transmit telescope emits short laser pulses at a rate of, say, 5 Hz. The departing laser pulses trigger an interval counter at a certain time (epoch of data); the laser pulses are reflected and the received laser light pulses are registered by a sensitive light-detecting device. The detector permits a stop signal to be sent to the interval counter. Utilising these round-trip time intervals (time-of-flight) and the speed of light, the range is half of this two-way range. Most SLR systems do require some form of human intervention to operate successfully. This is basically due to the fact that the combination of mechanical misalignment, clock offsets, instrumental and orbital biases as well as instrument- specific requirements creates a situation where the laser might be pointing slightly off target. The operator then has to make allowance for three offsets: along-track and cross-track biases, as well as time bias. These offsets are recorded on the computer and could serve as a starting point for subsequent satellite passes.

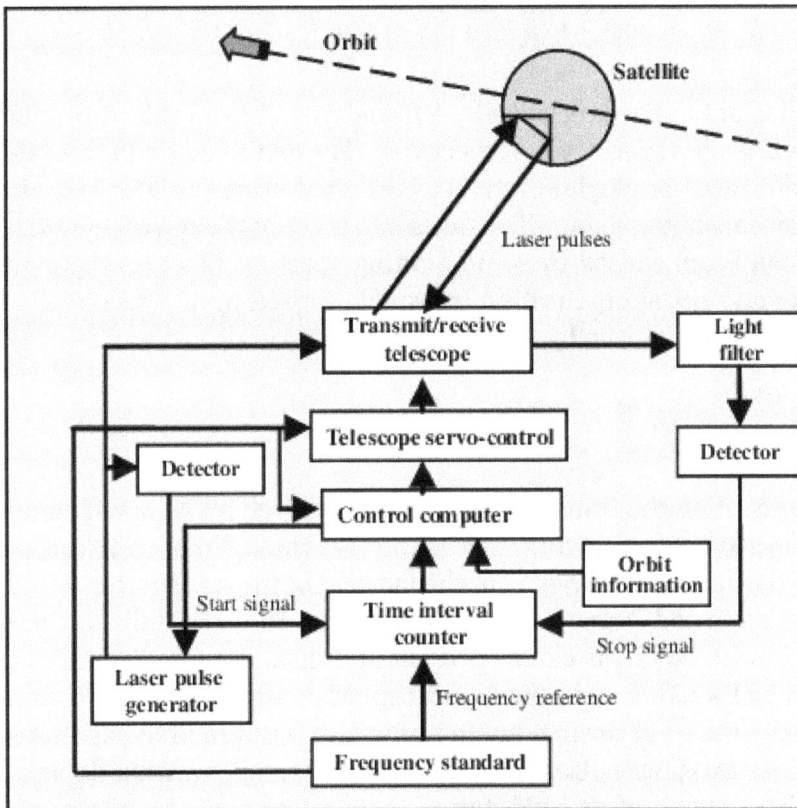

Basic components of a satellite laser ranging system, describing signal paths.

Apart from having to point in exactly the correct direction, other adequate conditions concerning laser power level, receiving telescope aperture and atmospheric conditions amongst others are required to receive successfully photons back from the satellite CCRs. The success of receiving returns can be estimated by the radar range equation, where the mean number of photoelectrons recorded by the SLR detector N_{pe} is given by,

$$N_{pe} = \eta_q \left(E_T \frac{\lambda}{hc} \right) \eta_t G_t \sigma_{sat} \left(\frac{1}{4\pi R^2} \right)^2 A_{R\eta r} \, T_a^2 T_c^2.$$

In the above equation, η_q is the detector quantum efficiency (the fraction of the total radiation incident that is actually detected), ET is the pulse power (average power divided by pulse repetition rate), λ is the wavelength of the laser, h is Planck's constant, c is the speed of light in a vacuum, η_t is the efficiency of the transmitter optics, G_t is the transmitter gain, the satellite optical cross-section is given by σsat and R is the slant range to the satellite. The effective area of the SLR telescope receiving aperture is A_R, η_r is the efficiency of the receive optics, T_a is the one-way atmospheric transmission and if cirrus cloud (thin, wispy cloud, occurring at altitudes >6 km, composed of ice crystals) is present, T_c is the one-way transmissivity of cirrus cloud. Slant range can be calculated using the equation,

$$R = -\left(R_E + h_l \right) \cos\theta_{zen} + \sqrt{\left(R_E + h_l \right)^2 \cos^2\theta_{zen} + 2R_E \left(h_s - h_l \right) h_s^2 - h_l^2}.$$

Here the radius of the Earth is given by RE and hl and hs are the altitudes of the station and satellite above sea level, respectively. The zenith angle of the satellite (complement of the elevation angle) θzen is as observed from the SLR station. A general expression for transmitter gain, which takes into account the effects of radial truncation of the Gaussian beam caused by some limiting aperture (such as the main transmitter primary) and central obscuration (normally caused by the secondary mirror of a Cassegrain telescope) is given by,

$$G_t = \frac{4\pi A_t}{\lambda^2} gt\left(\alpha_t, \beta, \gamma_t, X \right).$$

In the above equation the transmitting aperture is given by $A_t = \pi a2t$ and g_t (α_t, β, γ_t, X) depends on geometrical factors such as whether the collimating telescope is perfectly focused and whether the target is in the far field of the transmitter. More details are to be found in Klein and Degnan. Considering MOBLAS-6 and utilising these equations for a clear day with no cirrus clouds present and making some assumptions concerning good and bad weather conditions (atmospheric transmittance of 0.8 and 0.02, respectively), one finds that the maximum number of received photoelectrons could vary between 641 and 0.04 per pulse. This clearly illustrates the weather dependency of this technique. The location of an SLR station is therefore a critical factor; however, this matter seems to have carried less weight than other factors in the installation of several stations, e.g. collocation advantages with other space geodetic techniques as was the case with installing MOBLAS-6 at Hartebeesthoek Radio Astronomy Observatory in

South Africa. If there are no other compelling reasons, a site with minimum cloud and high atmospheric transparency will yield better results and more data than an inferior site and should be given preference.

In figure, a very important subsystem is the frequency standard; this unit as well as the time and frequency distribution throughout the other subsystems of the SLR is extremely important. An example will make this clear; the velocity of a satellite frequently used for different applications of SLR, LAGEOS, during a test analysis of a 1-day arc (~6.3 orbital revolutions) indicated a minimum velocity of 5,645 m/s and a maximum velocity of 5,806 m/s. This means that a timing precision of 1.7×10^{-7} or 0.17 μs is required to register the epoch of the observation. If the accuracy, i.e. how close the timing value is to the real value, is not exact, a small time bias can develop as the epoch of the observation can then be either too late or too early. Generally, SLR station time bias values are at the few microseconds level, and this is normally detected during post-analysis of the tracking data and corrected during the analysis. With regard to determining the range to the satellite, the velocity of light $\left(\sim 3 \times 10^8 \, \text{m/s} \right)$ gives one a two-way range precision of 0.15 mm/ps (ps = picosecond), so that in order to reach a few millimetres, a precision of at least 20 ps should be reached.

Currently, the objective of the SLR community is to reach millimetre accuracy in ranging, indicating that an improvement factor of 10 will be required. Therefore new systems, e.g. the 1-m SLR/LLR system being developed by HartRAO (South Africa) in collaboration with OCA (France), will require timing systems (event timer or interval counter) with 1- to 2-ps precision. This is a very demanding requirement; even so, several high-precision timing systems are being developed which approach these accuracies.

Range Model

This will describe how the real range to the satellite can be determined. Table gives a good overall view of some of the accuracies and precisions to be found in a modern SLR station. An average of 6- to 9-mm precision is achieved to a single CCR, 7- to 12-mm single-shot precision (1–3 mm for a normal point) for a geodetic satellite, with overall accuracy at 8–18 mm, i.e. ~1–2 cm. Normal points are made from a number of single shots, according to ILRS-prescribed guidelines; MOBLAS-6 averages about 66 shots (data points) per NP for LAGEOS. Limitations of space in this chapter preclude the discussion of all factors involved at great depth, but some of the issues involved in determining the range between the SLR station reference point and the satellite being tracked will become clear. The LAGEOS satellites will be used as example; principles involved will be more or less the same for other satellites.

An SLR station provides data in a specific format, as agreed to from time to time by the International Laser Ranging Service (ILRS) community. These data are uploaded to data centres and are consequently utilised to determine the range to a satellite; the ranges can then be used in a modelling process to estimate other parameters (Earth orientation, station position, grav ity coefficients, etc.). The ranging data basically consist of information such as satellite identification number, system-specific details such as wavelength of laser (532 nm in case of MOBLAS-6), calibrated system delay (two way in picoseconds), pass RMS (picoseconds) and epoch of laser firing in 0.1-μs units. Included is the main observable which is the two-way time-of-flight corrected for system delay in picoseconds. Other data are required to model the atmosphere and are also recorded, such as surface pressure, temperature and humidity. Raw ranges (taken at the transmit rate of the SLR, 5 Hz in the case of MOBLAS-6) are compressed to form a normal point.

Table: Laser ranging error budget for the French SLR stations at the turn of the century. These values are still representative of most modern SLR stations.

Origin	Precision (mm)	Accuracy (mm)
Laser	4–5	
pulse	1	
width	4–5	
Detector	3–6	
start	1–3	
return	3–5	
Timer	2–3	
Clock	1–2	
Calibration	1	2–6
geometry		1–2
electronic		1–4
Depend. (Az, El)	1–3	
Instrument	6–9	2–6
Atmosphere	3–5	5–8
pressure		1–2
temperature		1
humidity		4–5
Target signature		
LAGEOS (COM, etc.)	1–3	1–3
Single shot	7–12	
Normal point	1–3	8–18

The normal point (NP) data at a given epoch is taken here as the point of departure. This NP is converted easily to a normal point range in metres using the equation:

$$\mathrm{NPR}_i = \left(\frac{\mathrm{NPtof}_i}{1 \times 10^{12}} \times c \right) \Big/ 2\,(m),$$

where NPtof_i is the normal point time-of-flight (picoseconds) recorded at a certain epoch and c is the velocity of light (299,792,458.0 m/s). The range found in the above equation needs to be corrected by taking into account the effects of the atmosphere (Δa_i in equation below), the centre-of-mass correction (CoM) of the satellite (0.251 m for LAGEOS 2), SLR station range bias and a relativistic correction. A range equation below can then be written as:

$$NPR_i = \left(\frac{NPtof_i}{1\times10^{12}} \times c \right) \Big/ 2 - \Delta a_i + \Delta\,CoM_i - \Delta R_{b_i} - \Delta GR_i - \Delta\varepsilon_i,$$

where NPR_i (m) is the normal point range, i.e. the observed range. The centre-of mass correction is $\Delta\,CoM_i$ and the range bias, general relativity correction and a correction for unknown random errors are $\Delta\,CoM_i$, ΔR_{b_i} and $\Delta\varepsilon_i$, respectively. This observed range per normal point will be utilised to calculate the observed–calculated (O–C) residuals as part of the SLR data analysis process.

Laser use for Leveling in Surveying

Laser surveying instruments are primarily used to set elevation, grade or to plumb construction elements. Midwest Construction Equipment has created a list of commonly used laser surveying instruments, along with a brief description of their uses:

- Laser Levels – Laser levels point or rotate to create a reference point, line, or plane. When used outside, laser levels are almost always used with a receiver attached to a grade rod or mounted on heavy equipment for laser machine control. When used inside, a visible beam guides construction activity, but sometimes a receiver is used for large work areas. Some units have the ability to slope the plane on one or two axes to set a grade or slope. Laser levels used for machine control are more powerful.

- Laser Receivers – Receivers can attach to grade rods. Larger laser receivers are mounted on heavy equipment for machine control. Laser receivers are specific to the wavelength of light received. Newer laser receivers are universal for the wavelength of laser light received.

- Pipe Lasers – Used to set precise grade of pipes, drains or sewers, and are typically used with a pipe type- and size-specific target set.

- 3D Scanners – These devices use multi-point scanning to produce an accurate 3D model of any indoor or outdoor man-made or natural structure, mostly for project design support.

- Distance Meters – Whether a handheld device, part of a total station, or another type of surveying instrument, distance meters bounce laser energy off a surface or a target to measure distances with great accuracy.

- Laser Guided Machine Control – Machine Control is a broad category of application where the base technologies of laser, sonic and GNSS are applied individually or in combination to control heavy construction machinery. Single to multidimensional solutions are possible. A laser guided machine control uses receivers and sensors to either "indicate" by annunciation or they can "automate" to achieve the desired grade.

Laser Communication

A very useful and interesting application of laser is in the field of communications, which takes advantage of its wide bandwidth and narrow beam width over long distances. The laser beams can be created in a range of wavelengths from the ultraviolet to the infrared regions of the electromagnetic spectrum. The colour of the emitted light is relatively not important. The infrared region is preferred by the military, as it is more difficult to detect. The advent of semiconductor lasers has made possible the use of lasers for signal transmission. They are excited directly by electric cur-rent to yield a laser beam in the invisible infrared region.

A particular aspect of laser transmission, which makes it preferable to the ordinary radio waves for military purposes is the strict secrecy provided by the narrow beam width. Since no unwanted reception outside the narrow bundles of rays is possible, a high degree of secrecy can be maintained between two points, and thus, an interception-proof communication network can be realised. Besides, laser communication system is immune from jamming and from interference by spurious radio noise.

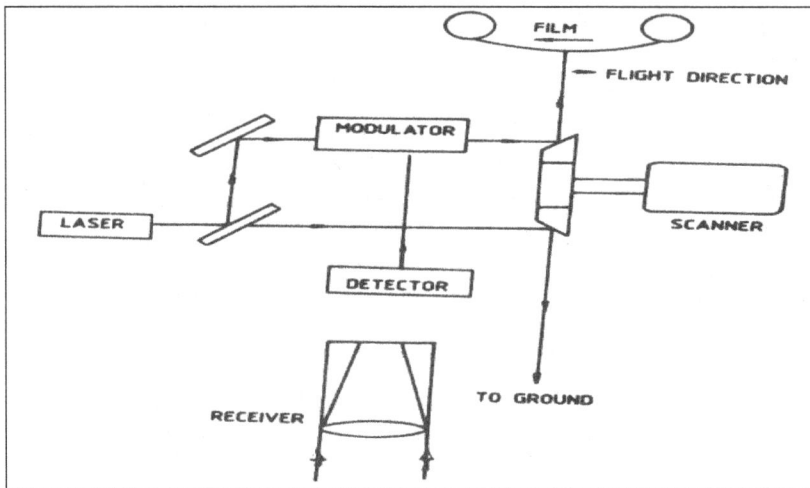

The optical laser has a great potential for use in long distance communication. Since the capacity of a communication channel is proportional to the frequency band width, at optical frequencies, the information carrying capacity is many times more than that is possible at lower frequencies. This and the fact that the laser is a generator of highly coherent beams which are powerful and sharply directed, make it ideally suited for communications.

In this regard, microwave technique offers direct competition to the laser as it has been perfected already to a high degree. Moreover, the optical frequency waves suffer a considerable disadvantage in case of atmospheric transmission since they are attenuated considerably by snow fog, and rain. Therefore, the laser communication through the atmospheric medium is effective only in clear weather conditions, with no obstacles interrupting the beam between the transmitting and the receiving stations.

Figure shows the principle behind the long distance communication system, which involves multiplexing the simultaneous transmission of different messages over the same pathway. For example, a channel for transmitting an individual human voice requires a frequency band extending from 2000 to 4000 cycles per second. For modulation of the signal without the addition of any noise, the carrier wave should be of a very narrow spectral width. This single frequency carrier wave is then successively modulated by a large number of voice signals to create a new composite single wave. With the help of special electrical networks, several broad communication bands are combined for simultaneous transmission over a single intensity pathway. On the other side of the line, a similar network separates the single signal into its component broad bands which are demodulated into individual telephone calls.

Thus considerable economy and efficiency in communication is achieved through multiplexing process. Since an individual communication channel requires the same bandwidth regardless of the region of the spectrum in which it is located, it is quite obvious from the above that the visible and near-infrared laser frequencies, which are about 1,000,000 times the frequency of the millimetre waves, offer great economy for communication.

For communication purposes, the laser beam is modulated by the signal. At the receiving station, the modulated beam is demodulated (detected) to separate the required signal from the laser beam (carrier). The output current, which varies with the intensity of the signal, is amplified and then fed to the speaker.

Most of the optical modulators devised so far are based on the variations in the refractive index of the substance used according to the signal wave. The continuous laser output from a laser passes through a polarisation modulator (KDP crystal). Ring electrodes are placed on the crystal and an electric field proportional to the signal wave is applied to the crystal, parallel to the axis. Due to the change in refractive index of the crystal, which follows the electric signal, there is change in polarisation of the light beam. As a result of this, the intensity of the light coming out of the analyser also changes, according to the signal.

The laser beam from a semiconductor laser can be directly modulated by varying the current through it, according to the signal. The demodulation of the laser beam can be accomplished in two ways: (i) by direct photodetectors and (ii) by photomixers.

Photomultiplier detectors are good to use in the visible and infrared regions. The method of demodulation by photoelectric detectors is shown in figure Silicon photodiodes, developed rapidly after the discovery of the laser, have a peak response at about 8500-9000 Å (one Å= 1×10^{-8} cm). This being the spectral region of the gallium arsenide lasers, the silicon photodiodes can be used as sensitive detectors in that region.

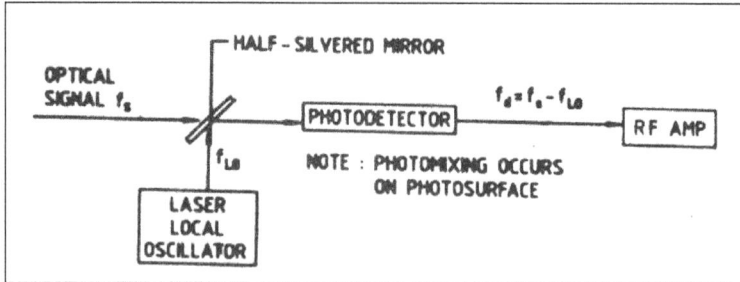

Demodulation by optical heterodyne detection is done by superposing on the incoming incident signal, a beam of light from an unmodulated laser called a local oscillator and allowing the resulting combined beam to fall on a photoemissive surface of the photo-detector. The electron current from the detector is modulated at a frequency equal to the difference between the signal and the local oscillator frequencies. When this is fed to an audio speaker, the input signal of the communication is reproduced.

Laser communication through open atmosphere is possible only when there is line of sight between the transmitter and the receiver and that too in good weather conditions. To circumvent these difficulties, laser communication through the medium of optical fibre has been achieved in recent years.

Advantages of Laser Systems in Communication

Laser communication systems offer many advantages over radio frequency (RF) systems. Most of the differences between laser communication and RF arise from the very large difference in the wavelengths. RF wavelengths are thousands of times longer than those at optical frequencies. This high ratio of wavelengths leads to some interesting differences in the two systems. First, the beam-width attainable with the laser communication system is narrower than that of the RF system by the same ratio at the same antenna diameters (the telescope of the laser communication system is frequently referred as an antenna). For a given transmitter power level, the laser beam is brighter at the receiver by the square of this ratio due to the very narrow beam that exits the transmit telescope. Taking advantage of this brighter beam or higher gain, permits the laser communication designer to come up with a system that has a much smaller antenna than the RF system and further, need transmit much less power than the RF system for the same receiver power. However since it is much harder to point, acquisition of the other satellite terminal is more difficult. Some advantages of laser communications over RF are smaller antenna size, lower weight, lower power and minimal integration impact on the satellite. Laser communication is capable of much higher data rates than RF.

The laser beam width can be made as narrow as the diffraction limit of the optic allows. This is given by beam width = 1.22 times the wavelength of light divided by the radius of the output beam aperture. The antennae gain is proportional to the reciprocal of the beam width squared. To achieve the potential diffraction limited beam width a single mode high beam quality laser source is required; together with very high quality optical components throughout the transmitting sub system. The possible antennae gain restricted not only by the laser source but also by the any of the optical elements. In order to communicate, adequate power must be received by the detector, to distinguish the signal from the noise. Laser power, transmitter, optical system losses, pointing system imperfections, transmitter and receiver antennae gains, receiver losses, receiver tracking losses are factors in establishing receiver power. The required optical power is determined by data rate, detector sensitivity, modulation format, noise and detection methods.

Laser Cutting

Lasers are used for many purposes. One way they are used is for cutting metal plates. On mild steel, stainless steel, and aluminum plate, the laser cutting process is highly accurate, yields excellent cut quality, has a very small kerf width and small heat affect zone, and makes it possible to cut very intricate shapes and small holes.

Most people already know that the word "LASER" is actually an acronym for Light Amplification by Stimulated Emission of Radiation. But how does light cut through a steel plate?

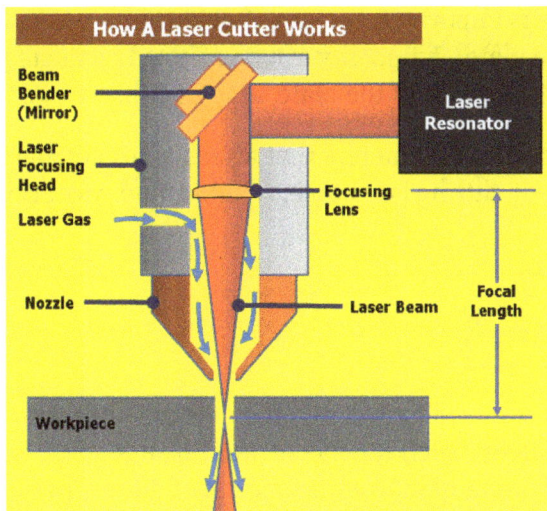

How it works?

The laser beam is a column of very high intensity light, of a single wavelength, or color.

In the case of a typical CO_2 laser, that wavelength is in the Infra-Red part of the light spectrum, so it is invisible to the human eye. The beam is only about 3/4 of an inch in diameter as it travels from the laser resonator, which creates the beam, through the machine's beam path. It may be bounced in different directions by a number of mirrors, or "beam benders", before it is finally focused onto the plate. The focused laser beam goes through the bore of a nozzle right before it hits the plate. Also flowing through that nozzle bore is a compressed gas, such as Oxygen or Nitrogen.

Focusing the laser beam can be done by a special lens, or by a curved mirror, and this takes place in the laser cutting head. The beam has to be precisely focused so that the shape of the focus spot and the density of the energy in that spot are perfectly round and consistent, and centered in the nozzle. By focusing the large beam down to a single pinpoint, the heat density at that spot is extreme. Think about using a magnifying glass to focus the sun's rays onto a leaf, and how that can start a fire. Now think about focusing 6 KWatts of energy into a single spot, and you can imagine how hot that spot will get.

The high power density results in rapid heating, melting and partial or complete vaporizing of the material. When cutting mild steel, the heat of the laser beam is enough to start a typical "oxy-fuel" burning process, and the laser cutting gas will be pure oxygen, just like an oxy-fuel torch. When cutting stainless steel or aluminum, the laser beam simply melts the material, and high pressure nitrogen is used to blow the molten metal out of the kerf.

On a CNC laser cutter, the laser cutting head is moved over the metal plate in the shape of the desired part, thus cutting the part out of the plate. A capacitive height control system maintains a very accurate distance between the end of the nozzle and the plate that is being cut. This distance is important, because it determines where the focal point is relative to the surface of the plate. Cut quality can be affected by raising or lowering the focal point from just above the surface of the plate, at the surface, or just below the surface.

There are many, many other parameters that affect cut quality as well, but when all are controlled properly, laser cutting is a stable, reliable, and very accurate cutting process.

Laser Welding

Laser welding is a process used to join together metals or thermoplastics using a laser beam to form a weld. Being such a concentrated heat source, in thin materials laser welding can be carried out at high welding speeds of metres per minute, and in thicker materials can produce narrow, deep welds between square-edged parts.

Laser welding operates in two fundamentally different modes: conduction limited welding and keyhole welding. The mode in which the laser beam will interact with the material it is welding will depend on the power density across the beam hitting the workpiece.

Laser Beam Welding

Laser Beam Welding is a fusion welding process in which two metal pieces are joined together by the use of laser. The laser beams are focused to the cavity between the two metal pieces to be joined. The laser beams have enough energy and when it strikes the metal pieces produce heat that melts the material from the two metal pieces and fills the cavity. After cooling a strong weld is formed between the two pieces.

It is a very efficient welding process and can be automated with robotics machinery easily. This welding technique is mostly used in the automotive industry.

Working Principle

It works on the principle that when electrons of an atom gets excited by absorbing some energy. And then after some time when it returns back to its ground state, it emits a photon of light. The concentration of this emitted photon increased by stimulated emission of radiation and we get a high energy concentrated laser beam. Light amplification by stimulated emission of radiation is called laser.

Main Parts

The main parts or equipment of laser beam welding are:

1. Laser Machine: It is a machine that is used to produce a laser for welding. The main components of the laser machine are shown.

2. Power Source: A high voltage power source is applied across the laser machine to produce a laser beam.

3. CAM: It is a computer-aided manufacturing in which the laser machine is integrated with the computers to perform the welding process. All the controlling action during the welding process by laser is done by CAM. It speeds up the welding process to a greater extent.

4. CAD: It is called as Computer-aided Design. It is used to design the job for welding. Here computers are used to design the workpiece and how the welding is performed on it.

5. Shielding Gas: A shielding gas may be used during the welding process in order to prevent the w/p from oxidation.

Types of Laser Used

1. Gas lasers: It uses mixtures of gases as a lasing medium to produce laser. Mixtures of gases such as nitrogen, helium, and co_2 are used as the lasing medium.

2. Solid-state laser: it uses several solid media such as synthetic ruby crystal (chromium in aluminum oxide), neodymium in glass (Nd: glass), and neodymium in yttrium aluminum garnet (Nd-YAG, most commonly used).

3. Fiber laser: The lasing medium in this type of laser is optical fiber itself.

Characteristics of Laser Beam Welding

1. The power density of laser beam welding is high. It is of the order 1 MW/cm^2. Because of this high energy density, it has small heat-affected zones. The rate of heating and cooling is high.

2. The laser beams produced are coherent (having the same phase) and monochromatic (i.e. having the same wavelength).

3. It is used to weld smaller sizes spot, but the spot sizes can vary from 2mm to 13 mm.

4. The depth of penetration of the LBW depends upon the amount of power supply and location of the focal point. It is proportional to the amount of power supply. When the focal point is kept slightly below the surface of the workpiece, the depth of penetration is maximized.

5. Pulsed or continuous laser beams are used for welding. Thin materials are weld by using millisecond-pulses and continuous laser beams are used for deep welds.

6. It is a versatile process because it is capable of welding carbon steels, stainless steel, HSLA steel, aluminum, and titanium. Due to the high cooling rate, the problem of cracking is there when welding high-carbon steels.

7. It produces high-quality weld.

8. This welding process is most popular in the automotive industry.

Working

1. First, the setup of welding machine at the desired location (in between the two metal pieces to be joined) is done.

2. After setup, a high voltage power supply is applied to the laser machine. This starts the flash lamps of the machine and it emits light photons. The energy of the light photon is absorbed by the atoms of ruby crystal and electrons get excited to their higher energy level. When they return back to their ground state (lower Energy state) they emit a photon of light. This light photon again stimulates the excited electrons of the atom and produces two photons. This process keeps continue and we get a concentrated laser beam.

3. This high concentrated laser beam is focused to the desired location for the welding of the multiple pieces together. Lens is used to focus the laser to the area where welding is needed. CAM is used to control the motion of the laser and workpiece table during the welding process.

4. As the laser beam strikes the cavity between the two metal pieces to be joined, it melts the base metal from both the pieces and fuses them together. After solidification, we get a strong weld.

5. This is how a laser Beam Welding Works.

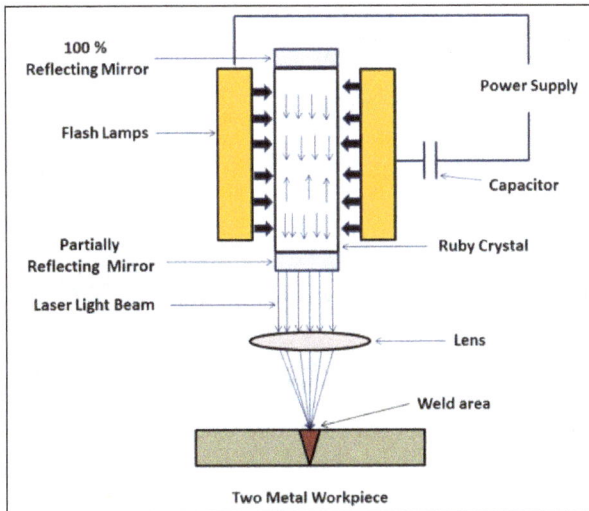

Two Metal Workpiece

Advantages

1. It produces high weld quality.

2. LBW can be easily automated with robotic machinery for large volume production.

3. No electrode is required.

4. No tool wears because it is a non-contact process.

5. The time taken for welding thick section is reduced.

6. It is capable of welding in those areas which are not easily accessible.

7. It has the ability to weld metals with dissimilar physical properties.

8. It can be weld through air and no vacuum is required.

9. X-Ray shielding is not required as it does not produce any X-Rays.

10. It can be focused on small areas for welding. This is because of its narrower beam of high energy.

11. A wide variety of materials can be welded by using laser beam welding.

12. It produces a weld of aspect ratio (i.e. depth to width ratio) of 10:1.

Disadvantages

1. The initial cost is high. The equipment used in LBW has a high cost.

2. High maintenance cost.

3. Due to the rapid rate of cooling, cracks may be produced in some metals.

4. High skilled labor is required to operate LBW.

5. The welding thickness is limited to 19 mm.

6. The energy conversion efficiency in LBW is very low. It is usually below 10%.

Laser Scanning for Building Design and Construction

Laser scanning is a method of collecting surface data using a laser scanner which captures the precise distance of densely-scanned points over a given object at rapid speed. The process is commonly referred to as a point cloud survey or as light detection and ranging (LIDAR, a combination of the words 'light' and 'radar'). It can be used to generate 3D imagery that can be converted for use in 3D computer aided design (CAD) modelling or building information modelling (BIM).

Instrumentation comprises high-speed lasers with an integrated camera using colour coding mounted on a tripod. Typically such instruments operate up to a range of 180 metres and at speeds of up to 990,000 points per second.

Accuracy

Laser scanning has proved to be much quicker, more accurate and cheaper than conventional survey measurement. The accuracy of the process depends on the steadiness of the instrument base and the distance from the object. Close range objects achieve sub millimetre accuracy. For normal terrestrial survey work + or − 2mm per 100m is a good guide to accuracy. Greater distances of 2 kms may be accurate to + or − 50mm.

Use in Property and Construction

Laser scanning provides a robust method for surveying inaccessible surfaces as well as complex geometry. All the major providers of CAD 3D modelling and BIM software have built compatibility that allows their systems to import the point cloud data into 3D visual graphic material.

The use of helicopters and drones with laser scanning has become a recognised method capturing the exact detail of topography, existing structures and townscapes. LIDAR has also come become invaluable for surveying existing properties for retrofitting and refurbishment.

LIDAR has been extensively used for surveys from moving rail bogeys and road vehicles. The instruments can operate at night when the targeted surfaces are less obstructed by people although such imagery will be seen in black and white only. Night time operation can produce greater accuracy.

The Future

It is possible in the future that LIDAR technology will be used in conjunction with 3D printers to manufacture and replace building components, resulting in savings in the storage of spares for maintenance.

During construction, progress photographs may become a thing of the past. LIDAR will give instant and accurate 3D visual comparisons between anticipated planned progress and actual progress. This in turn might be linked to interim payments for contractors.

LIDAR may become the most effective and most accurate way to record as-built information.

Laser Printers

Laser printers are one of the many types of printers available on the market today. Instead of using ink, lasers use toner to replicate the object, and then transfer it to paper. Businesses use laser printers almost exclusively because they have the reputation of being reliable while making a quality print product. Some common uses for laser printers include printing company stationery, making labels, and creating company fliers and brochures.

Types

Monochrome laser printers only print black text. Companies use monochrome laser printers to make letters, spreadsheets and other text documents. Color laser printers add color to the prints by using four toners: black, red, cyan and yellow. One uses color laser printers in graphic design, such as adding a logo to letterhead or making a slick brochure. Large format printers mean the printer prints sizes larger than 8 by 11. Large format laser printers may be monochrome or color. Companies use large format laser printers to make highway signs and billboards.

Other Uses

Use a laser printer for mass production. Laser prints come out of the printer dry. Some inkjets come out wet and if one does not allow the paper to dry, the page will smear or "bleed through," ruining the print product. Use a color laser printer with high resolution for sharp, clear photos. Laser printers with high resolution result in clearer, more life-like pictures. High resolution is a technical term that refers to the clarity of the printer product. Resolution is measured in dots per inch, or DPI, and DPI is listed as one of the features of any printer. If you plan to make photos or graphics with your laser printer pick one with the highest DPI available.

Benefits

Laser printers are not messy. Since the toner is a dry powder substance, there is no chance of spilling ink on papers, desks or you. Laser printers turn out high-resolution products. This means the print is clear and readable. Laser printers turn out a dry product fast. The average life of a laser printer is 10 years or more.

Home Uses

At one time, laser printers were almost exclusively used by businesses because the price put them out of reach of most consumers. Today, laser printer companies make low cost laser printers intended for home use. These printers are not quite as rugged as their industrial counterparts, but they do have all of the properties that make laser printers a desirable choice. Find laser printers for home use at local electronic stores. Choose a laser printer to make greeting cards, personal letters and resumes.

Laser printers cost substantially more than inkjet printers. Toner cartridges also cost more than an ink cartridge. Lasers use heat in the production of printed materials, so one must make sure the area is well ventilated or the printer may burn out or smell. Laser printers are large so they may not fit in tight quarters.

Medical uses of Laser

Lasers in Dermatology

Lasers were introduced in the specialty of dermatology in the mid-1960s. Because of the accessibility of the skin to examination and study by lasers, dermatologists play an extremely important role in defining the clinical usefulness and limitations of laser systems. They have also helped to define the specificity of laser–tissue interaction that further improved the usefulness of these devices. When the absorptive characteristics of a targeted tissue are precisely matched with an ideal laser wavelength, the maximal specificity of the laser–tissue interaction can occur. The optical property of the skin is an important determinant for the selectivity of laser effects. There are two main chromophores in the skin: oxygenated haemoglobin with three absorption peaks at 418, 542 and 577 nm; and melanin which has a very broad range of absorption. In addition, water is the key component of the skin tissue that can affect the quality of the thermal effects ranging from structural changes of protein at temperatures of 42–45 °C, to coagulation at 50–60 °C and vaporization at above 100 °C.

There are a large number of lasers used for various types of dermatological indications including visible wavelengths of pulsed dye, ruby, KTP, diode, alexandrite and infrared wavelengths of CO_2, Nd: YAG. The major effect of lasers on skin tissue is photothermolysis and the common skin indications treated with lasers are vascular lesions, benign and malignant tumours, infectious lesions, pigmented lesions and tattoos and a number of cosmetic conditions. Generally, lasers used correctly causes limited side-effects with little discomfort, no risk of infection and no scarring.

One of the most common cutaneous vascular lesions treated with lasers is the port-wine stain (PWS). PWSs with an incidence of about 0.4% of newborns are benign vascular birthmarks consisting of superficial and deep dilated capillaries in the skin resulting in a reddish to purplish discoloration. PWS can cause significant psychological trauma and a reduction in the quality of life. Treatment of PWS includes ionizing radiation, cryosurgery, skin grafting, but lasers provide the treatment of choice for most patients today. Being absorbed by intravascular oxyhaemoglobin within the visible-light range a laser can induce photothermolysis to destroy the diseased blood vessels selectively without affecting the surrounding tissues and causing scarring. Various types of lasers have been used for PWS, but the flashlamp-pumped pulsed dye laser (PDL) is the treatment of choice with the wavelengths from 585 to 600 nm. In addition, the KTP laser at the shorter wavelength of 532 nm is still used for the PWS treatment with the advantage of causing less purpura. PDT is now also applied for PWS. Unlike PDL-induced coagulation, PDT uses continuous low light irradiation with no photothermal effect to activate a photosensitizer that has accumulated in diseased vessels and thus selectively destroys the vessels.

Although thermal laser therapy is used for the treatment of skin tumours, PDT with ALA or its methylester has recently been applied for the treatment of cutaneous premalignant and malignant lesions. The principle of ALA-PDT is that in the initial step of the heme biosynthetic pathway in cells ALA is formed from glycine and succinyl CoA. The last step is that the rate-limiting enzyme, ferrochelatase, catalyzes the insertion of ferrous iron into PpIX (a potent photosensitizer) in the mitochondria. By adding exogenous ALA, the naturally occurring PpIX accumulates with a high degree of selectivity in tumours possibly because of limited capacity and/or low activity of ferrochelatase. Such a selectivity has therefore been exploited for its application in photodetection and PDT of tumour as an alternative to administration of exogenous photosensitizers.

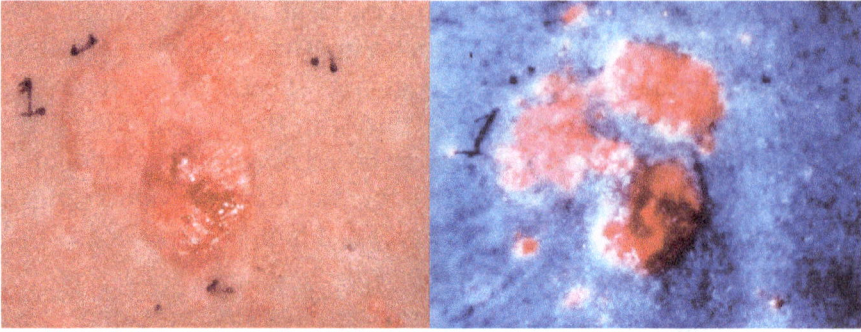

Highly selective ALA methylester (Metvix)-induced
PpIX fluorescence (right) in human skin basal cell carcinoma.

The benefits of ALA-based PDT include reduced skin photosensitivity (1 or 2 days compared with 1 or 2 months with some other photosensitizers), easy administration (topical or oral) and repeat use if necessary. PDT with topically applied ALA or its methyl ester has already been approved by regulatory agencies of many governments for the treatment of actinic keratosis and basal cell carcinoma.

Q-switched lasers with short-pulsed and high intensity are used to bleach the dyes in unwanted tattoos and fade the pigments in age spots and moles. The 308 nm excimer laser is effective to clear psoriasis. In addition, lasers are applied for resurfacing, hair removal and wound healing.

Lasers in Ophthalmology

No field has seen greater accomplishments with lasers than ophthalmology. The advantage in this field is the ability of a laser beam to enter the eye without causing injury. In the 1940s, Gerd Meyer-Schwickerath, a German ophthalmologist, pioneered to use the energy of the sun to 'weld' the retina to the underlying epithelium. This was the first eye surgery with light coagulation of the retina. Ophthalmology was perhaps the first subspecialty in medicine to use laser light to treat patients. Immediately after the invention of the first laser in 1960, Campbell used a confocal laser transmission system for the first retinal laser coagulation in 1961. Today, lasers are indispensable for effective and minimally invasive microsurgery of the eye. The focusing system of the cornea and lens brings laser beams to a sharp focus inside the eye. This actually carries a risk of injury, but also has considerable therapeutic possibilities. Generally, laser energy has four different effects of light–tissue interaction on the eye: photodisruption, photoablation, photocoagulation and photochemical reactions.

Photodisruption uses the mechanical energy of a laser to create microexplosions, expanded plasma formation, acoustic waves and cavitations in order to cut intraocular structures with minimal thermal damage. The Nd:YAG laser, that can induce a photodisruptive effect to make tiny holes, is applied for iridotomy in the pupillary-block glaucoma caused by increased pressure in the eye, and also for lens capsulotomies.

Photoablation involves minimal thermal damage. Hyper- opia (longsightedness), myopia (shortsightedness) and astigmatism can be corrected by photorefractive keratectomy (PRK) with non-thermal excimer laser ablation in the far ultraviolet. The laser-assisted *in situ* keratomileusis (LASIK) is a commontype of the PRK and is done in the stromal bed by creating a plano-hinged flap with a keratome. Myopia up to 12 dioptres and hyperopia up to 5 dioptres, each with a certain degree of astigmatism, can be treated effectively in this way. Patients choose LASIK as an alternative to wearing corrective eyeglasses or contact lens. In addition, the laser ablation can be used as phototherapeutic keratectomy to remove superficial corneal opacifications in scars and dystrophies and to close the epithelium in non-infectious corneal ulcers.

Photocoagulation uses the thermal energy of a laser to seal leaking blood vessels. The argon laser is a standard modality for retinal coagulation of diseases such as diabetic retinopathy and vascular diseases. In the treatment of glaucoma, focal coagulation can be induced within the trabecular meshwork of the anterior-chamber angle with the argon laser using a high energy to cause the outflow of the aqueous humour, thus lowering the intraocular pressure. The excimer and Er-YAG lasers can also be used to produce a few tiny holes in the trabecular meshwork. In vascular and pigmented eyelid lesions including tumours CO_2 and Er-YAG lasers are used for tissue coagulation to reduce bleeding, making the surgical field easy to survey.

The photochemical effect generated by PDT with verteporfin and a low-energy laser is used to treat choroidal neovascularization (CNV) secondary to age-related macular degeneration (AMD). Until recently, the only treatment proven beneficial was thermal laser photocoagulation. However, the thermal laser damages the retina overlying the CNV, making it problematic for lesions involving the foveal centre. The photochemical application has shown to reduce the risk of moderate and severe vision loss in patients with predominantly classic subfoveal CNV secondary to AMD.

A scanning laser ophthalmoscope is a diagnostic tool for eye disorders. The principle is that a focused low-level laser beam is used to scan the fundus of the eye to provide a digital and fluorescent image of the retina, in which the depth of the optic cup and the thickness of the retinal nerve fibre layer can be measured.

Lasers in Dentistry

The common lasers used today in different fields of clinical dentistry are argon, KTP, HeNe, diode, Nd : YAG, ErCr : YSGG, Er : YAG and CO_2. Among them He–Ne laser and diode laser (632 nm) with a low power are applied for photodetection and PDT. These fields mainly include periodontology, endodontics, hard tissue applications, soft tissue surgery and esthetic dentistry.

Periodontal diseases are among the most prevalent infectious diseases, with 75% of people aged between 35–44 and 95% of the population over 65-year olds being affected. Most pathogenic bacteria in the oral cavity are present as complex aggregates (biofilms) on the surfaces of the teeth (known as dental plaques). The accumulation of these bacterial biofilms on the tooth surface above the level of the gingival margin causes the progressive dissolution of enamel and the underlying dentine to induce caries.

Treatments of carious lesions include the elimination of infected dentine by drilling and restoration of the tooth by filling with a variety of materials. However, instead of the removal of the infecting microorganisms by drilling, a more attractive alternative approach is to kill the bacteria in situ. Recent studies have shown that the bacteria present in dental plaques and caries are susceptible to lethal photosensitisation of the photosensitizer Toluidine Blue O or aluminium disulfonated phthalocyanine to a laser. This would help not only to prevent dental caries but also to eliminate infected dentine. Periodontitis is a chronic inflammation due to the accumulation of bacterial biofilms on the root surface of the tooth below the level of the gum margin (subgingival plaque). The bacterial biofilm induces destruction of the periodontal ligament and a gap between the tooth and the gum (periodontal pocket), resulting in the loosening and eventually loss of the tooth. The physical removal of the subgingival biofilms combined with topical application of antimicrobial agents to the periodontal pocket is the conventional therapy. However, it is difficult to maintain therapeutic concentrations of the antimicrobial agents due to the dilution by the saliva and gingival crevicular fluid. PDT is feasible, as a photosensitizer can be directly applied to the periodontal pocket followed by light irradiation either through the thin gingival tissues or via an optical fibre placed into the pocket. A number of studies have shown that PDT can effectively kill periodontopathogenic bacteria.

In the case of infected root canal system the routine endodontic treatment procedures are to mechanically enlarge the canals followed by irrigation with an antibiotic agent and filling of the affected space. However, the complex anatomy of the root canals makes it difficult or virtually impossible for the routine procedures to completely remove the bacteria. PDT may offer an alternative to disinfect and sterilize the canals due to the easy access of a photosensitizer and light.

Overall, the advantages of PDT over conventional antimicrobial agents are (a) a faster process of killing bacteria (seconds to minutes), (b) can be repeated without inducing drug-resistance and (c) sparing surrounding tissues. Argon and diode lasers with a low power are used to excite hydroxyapatite and bacterial by-products for fluorescence detection and quantification of incipient occlusal and dental carious lesions in pits and fissures. This fluorescence technique has greater sensitivity than conventional visual and tactile methods. Since laser exposure lacks the risks of ionizing radiation, this allows its frequent use for monitoring dental lesions. Apparently, the best is the combination of a fluorescence detection system with a therapeutic laser to allow diagnosis and treatment of dental diseases in a single device.

ErCr : YSGG and Er : YAG lasers, operating at the wavelengths of 2780 nm and 2940 nm, respectively, are used for dental hard tissues including caries removal and cavity preparation without significant thermal effects, collateral damage to tooth structure and patient discomfort. This is because normal dental enamel contains sufficient water and these Er-based laser wavelengths correspond to the peak absorption of water, so that they can achieve effective ablation at temperatures well below the melting and vaporization temperatures of the enamel. For example, the most uncomfortable component in dental treatment is probably the drill (handpiece). Er : YAG and CrEr : YSGG lasers can drill into the enamel and dentin, and may replace the dental drill with the advantages of minimal pain and no noise or vibration. The CO_2 laser is also highly absorbed by dental hard tissues, but it is not suitable for drilling and cutting enamel and dentine because its deeper thermal absorption can damage the dental pulp.

Numerous surgical procedures for soft tissues can be performed with lasers including gingivectomy, gingivoplasty, removal of gingival pigmentation, frenectomy, vestibuloplasty, aphthous removal, tumour excision, etc. Argon, KTP, Nd : YAG, ErCr : YSGG and Er : YAG lasers are usually used for minor soft tissue surgery, whereas CO_2 lasers are suitable for major soft tissue procedures. This is due to the efficient absorption of many commonly used dental lasers by water and haemoglobin in oral tissues. The main advantages of laser soft tissue surgery are reduced bleeding and less pain. Lasers can also be used for aesthetic dentistry such as tooth whitening. Diode (810–980 nm) and CO_2 (10 600 nm) lasers are used to induce photothermal bleaching; while argon (514.5 nm) and KTP (532 nm) lasers are used for photochemical bleaching due to the fact that their wavelengths fit well with the maximal absorption (510–540 nm) of those chelate compounds formed between apatites, porphyrins and tetracycline compounds.

Lasers in Otolaryngology

Experimentally, the first laser in otolaryngology was the ruby laser that was used on the inner ear of pigeons in 1965. A Nd : YAG laser was then applied to the otosclerotic footplate in 1967 and an argon laser to the otic capsule in 1972. Today, laser technology is used in virtually all areas of the ear, nose and throat specialty. Although a number of lasers including CO_2, argon, KTP, Ho : YAG, Nd : YAG, diode lasers are available today, only the CO_2 laser is in routine use as a workhorse laser in otolaryngology because of its excellent cutting properties with little lateral tissue damage. The major advantage of the laser over a scalpel is, for example, that any damage to the voice box by larynx surgery can severely impair speech; while the laser enables lesions to be vaporized without additional damage to the larynx. Lasers have also been used to treat snoring by trimming away part of the uvula.

The CO_2 laser, with the help of a microlaryngoscope as well as a delivery system, offers a distinct advantage in the management of polyps, nodules, cysts, leukoplakia, subglottic stenosis, webs, capillary hemangiomas, etc in laryngology. Treatment of these conditions needs careful attention to preserve the phonatory mechanism, as the vocal ligament may be involved. Micro-point manipulators on the newer generation of CO_2 lasers provide an excellent precision in the management of certain conditions such as haemorrhagic polyp and Reinke's oedema. With an appropriate power setting in a pulsed mode, the capillaries are sealed off during surgery, facilitating the incision and dissection. The bilateral abductor paralysis, a neuromuscular disorder causing dyspnoea, can be treated with a laser to obtain satisfactory airways on phonation. Treatment of laryngeal tumours with a laser has been the subject of much debate and controversy, although lasers are the method of choice for treating recurrent respiratory papillomatosis.

In rhinology, almost any lesion in the nasal cavity can be treated with a laser. The nasal mucosa is extremely vascular and conventional surgical procedures (punch and grasping forceps) in the nose cause profuse bleeding, which obscures the view and makes the procedures become somewhat 'blind'. The relatively bloodless field offered by a laser and an endoscopic application device provides the surgical procedure with a better visual control of the surgical field and surrounding structures. Generally, effective preoperative decongestion of the nasal mucosa with a laser is almost universally useful. The benefits of nasal laser management are dependent on the nature of the lesion, laser energy as well as wavelength, surgical technique and the operator's experience.

In otology, laser applications for ear surgery are still very limited. The operative procedure on the ear largely involves gross bone removal. The energy provided by a laser may be utilized to undertake certain steps of the procedure where an extreme degree of finesse is required or where conventional procedures can produce unwanted effects. At a low intensity, argon laser radiation penetrates through the bone without altering it, and may thus be used to devitalise extensions of cholesteatoma within the cell spaces of the mastoid bone. At a high energy, it vaporizes the bone, and has been used to undertake fenestration of the footplate in otosclerosis.

PDT has been tried to treat head and neck cancer for more than two decades, and Foscan as a photosensitizer has recently been approved for such PDT treatment.

Lasers in Gastroenterology

Gastroenterology was one of the earliest specialties to examine the use of lasers in the early 1970s for the arrest of gastrointestinal haemorrhage. Because of the developments in gastrointestinal endoscopic techniques and the optic fibres inserted through the instrument channels of endoscopes, laser light can be easily delivered to the upper and lower gastrointestinal tracts in a safe and relatively non-invasive fashion.

The most important laser used today in gastroenterology is Nd : YAG. Short shots from this laser obtain good haemostasis due to thermal contraction of soft tissue. Longer shots at high powers can vaporize tissue and coagulate the underlying layers for effective debulking of advanced tumours; while those at much lower powers can coagulate a larger volume of tissue without vaporization. Thermal lasers in current practice are used for palliation of advanced, inoperable cancers of the upper and lower gastrointestinal tract. Under direct vision with the laser fibre held away from the surface of the target, nodules of exophytic tumour can be vaporized and the underlying tumour coagulated either to relieve obstruction or to reduce blood loss. Laser therapy can improve dysphagia in patients with cancers of the esophagus and gastric cardia, but several treatments and the introduction of expanding stents are often needed to achieve optimum recanalization.

The tip of the laser fibre can also be directly inserted into a targeted tissue with a much lower power to gently 'cook' the targeted area over a period of several minutes. This is called interstitial laser photocoagulation (ILP). ILP can be used for the percutaneous treatment of small hepatic metastases under the guidance of ultrasound, computerised tomography or magnetic resonance imaging in patients who are unsuitable for partial hepatectomy. Because of the simple and cheap sclerotherapy for haemostasis, lasers are not often used for the control of haemorrhage today, but they still play an important role in controlling blood loss in vascular lesions such as hereditary telangiectasia and angiodysplasia. The Q- switched or pulsed Nd : YAG laser is often used for fragmenting biliary stones.

Perhaps the most important new applications of lasers in gastroenterology are PDT being developed for the treatment of dysplasia in Barrett's esophagus and small tumours in the gastrointestinal tract. Potential applications of PDT also include tumours of the pancreas and bile duct. The major advantage of PDT over thermal laser therapy is the selective effect on the mucosal layer of the gastrointestinal tract with little damage to underlying connective and muscular tissues, thus leading to minimal risk of perforation.

Lasers in Urology

Since the first report on the use of a ruby laser to fragment urinary calculi in 1968, this technology has been extensively applied by urologists. Laser lithotripsy in the 1980s and now laser prostatectomy have dominated laser usage in urology. Because of improvements in endourology, nephroscopy, perinephroscopy, laparoscopy, pelviscopy and retroperitoneoscopy, almost all urologic organs have become visually accessible via fibreoptic laser delivery. Laser applications in urology are based on several mechanisms for laser–tissue interactions including photomechanical, photothermal and photochemical effects.

Clinical applications of laser lithotripsy began in the mid-1980s with the introduction of the pulsed dye laser. The laser operates with a short pulse of 1 µs at a wavelength of 504 nm. This is based on a photomechanical mechanism. The optical fibre is placed in direct contact with the stone, and a short laser pulse is delivered, resulting in a shockwave at the stone surface to fragment most types of stone. However, due to its large size and high cost as well as maintenance, the dye laser has been replaced with the Ho: YAG laser. Unlike the dye laser, the Ho: YAG laser operates with a long pulse of approximately 500 µs, generating a photothermal mechanism of stone fragmentation with chemical decomposition of the irradiated calculus components.

The Ho: YAG laser wavelength of 2120 nm is strongly absorbed by water in the tissue, allowing the laser to cut and coagulate tissue with a minimum zone of 0.3–0.4 mm. This property makes the laser also ideal for a range of soft tissue procedures including benign prostate hyperplasia (BPH) and bladder tumour as well as strictures. BPH, or enlargement of the prostate gland, is a common benign disease that occurs with increasing age in the male population. Recently, the frequency- doubled KTP and Nd : YAG lasers with a high power have been introduced for vaporization of BPH. The KTP laser wavelength of 532 nm is strongly absorbed by haemoglobin and therefore provides excellent haemostatic properties during vaporization and coagulation of the prostate, although with a larger thermal damage zone (1–2 mm) than the Ho : YAG laser. Thus, the KTP laser represents a higher power, less expensive, more reliable solid-state alternative.

Other common applications of lasers in urology include Ho : YAG laser incision of urethral strictures caused by surgery-induced tissue trauma, Ho : YAG laser ablation of superficial bladder transitional cell carcinoma, Nd : YAG and CO_2 laser ablation of penile carcinoma and Nd : YAG and Ho : YAG laser incision of ureteroceles.

Although PDT is still under clinical development, it has shown promising results in the treatment of bladder cancer and small prostate and penile cancers with several photosensitizers including Photofrin and Tookad. Hexvix, a porphyrin precursor, has recently been approved by the EU for the photodetection of bladder cancer.

Lasers in Gynaecology

Since the CO_2 laser was first used in gynaecology more than 20 years ago, a number of other lasers have been used in this field such as Nd : YAG, KTP, dye and diode lasers. As more gynaecological procedures move to laparoscopic, colposcopic and hysteroscopic surgery, the use of lasers continues to increase. Advances in laser technique have improved precision and minimized thermal damage. These make it possible to use a laser for incision, excision, resection, ablation, vaporization, coagulation and haemostasis of soft tissues in the field of gynaecology. With colposcopy and CO_2 laser ablation, for example, condyloma, leukoplakia and high-grade cervical, vulvar and vaginal intraepithelial neoplasia (CIN, VIN and VAIN) can be treated. A number of reports have shown this laser to be effective at treating VIN with a success rate of approximately 85%. By laparoscopy/hysteroscopy and photovaporizing diseased or excess tissue, CO_2 and KTP lasers have also been used to treat ectopic pregnancies, dysmenorrhea, endometriosis, ovarian cysts, etc. In addition, lasers have been employed to perform hysterectomies and to reconstruct damaged or blocked fallopian tubes, thus allowing infertile women to conceive. The major advantage of laser treatment is the eradication of any volume of diseased epithelium, with negligible risk of scarring. Recently, fluorescent-dyed tissue samples with porphyrin precursors have been shown to reveal abnormal cells when excited by laser light, leading to early detection of cancer such as cervical cancer.

Lasers in the Cardiovascular System

By fibreoptics laser radiation can be transmitted to anywhere in the cardiovascular system. This makes laser-based modalities attractive, and today laser technology is available for the treatment of artery disease, ventricular and superventricular arrhythmias, hypertrophic cardiomyopathy and congenital heart disease. The first laser application to treat cardiovascular diseases was made by McGuff *et al* in 1963. They used a laser to vaporize atherosclerotic plaques. Choy et al used argon laser radiation to treat thrombosis in animals, and then performed the first clinical coronary argon laser angioplasty with a bare fibre in 1983. In contrast to balloon angioplasty where plaque material is fractured, compressed, or displaced, laser angioplasty vaporizes the plaque material and thereby has a high success rate for treating chronic coronary artery occlusions. This approach is often referred to as laser thermal angioplasty, since the tips of optical fibres can convert the laser light energy into heat energy to achieve recanalization by mechanical compression and tissue vaporization of the plaque material. The results reported for excimer laser angioplasty show success rates of 82–85% with major complication rates of only 5–7%. Failures of laser angioplasty occur often due to the inability to advance the catheter to the lesion because of prestenotic vessel tortuosity.

A laser can also ablate thrombi and emboli by photovaporization. This is because light absorption by haemoglobin in thrombi is larger than that by vascular tissue at 482 nm, thus providing a degree of selective laser ablation of the thrombi without damage to vascular walls. Gregory et al reported selective laser thrombolysis of coronary artery thrombi in 17 out of 18 patients with significantly improved coronary blood flow.

Transmyocardial laser revascularization is a technique for the treatment of patients with chronic angina pectoris. This procedure employs a laser to create multiple transmyocardial channels in the ischemic areas. The procedure can also be performed by percutaneous myocardial laser revascularization with a less-invasive advantage. Clinical trials have shown a significant improvement in angina class, but it is too early to draw any clear conclusion, as the techniques have not yet indicated any significant increase in survival and myocardial functions.

Laser ablation has been studied in the treatment of ventricular and supraventricular arrhythmias. Saksena et al and Isner et al used argon and CO_2 lasers to achieve superficial vaporization of endocardial tissue responsible for ventricular tachycardia. Compared with argon and CO_2 lasers for tissue vaporization, the Nd : YAG laser has been proposed for antiarrhythmic therapy by in situ photocoagulation of the inciting focus. One of the advantages of Nd : YAG laser photocoagulation is that the treated tissues are left intact, preserving structural integrity of the myocardium. The major benefit in contrast to cryoablation is that Nd : YAG laser photocoagulation can be performed on the normothermic beating heart during ventricular tachycardia.

Although laser technology has been evaluated for the treatment of a number of cardiovascular disorders for several decades, it has not performed as well as hoped.

The procedure may create dangerous blood clots and perforation of vascular wall due to laser ablation. Catheters have been reported to cause mechanical trauma and are also too stiff to pass through convoluted blood vessels. Further research on both laser technology and its cardiovascular application may find the laser valuable in the cardiovascular system.

Since initial experimental studies on the effects of ruby, argon and CO_2 lasers on the central nervous system (CNS) of rats, dogs and pigs in the period 1965–1970, dramatic progress has been made with respect to both technical and surgical applications in brain and nervous tissues. The major effect of a laser on neural tissue is thermal. Today, the CO_2 and Nd : YAG lasers, perhaps also new high power diode lasers, are effective in the treatment of CNS tumours and vascular malformations. Generally, laser-induced tissue vaporization is suited for resection of intracerebral, extra-axial, skull base and spinal tumours including acoustic neuromas, pituitary tumours, spinal cord neuromas, intracerebral gliomas and metastases; and also for dissection of intracranial, spinal cord and intra-orbital meningiomas; while laser-induced tissue coagulation is used for the resection of vascular malformations such as arteriovenous malformations and cavernomas. The benefits of such laser therapy are due to the accurate and non- contact instruments that can reduce surgical brain trauma.

It should be pointed out that normal and abnormal brain tissues may have very different optical properties. For example, most brain tumours (meningioma, neuromas, high- grade gliomas, metastases) are highly vascularised with a high haemoglobin level and thus demonstrate high absorption coefficients in the 400–800 nm wavelength band. This indicates that argon and diode lasers may be effective for these types of tumours.

Stereotactic techniques are more often used for neurosurgery because of their smaller openings, reduced brain injury, decreased morbidity and a shorter postoperative period. In fact, any fibreoptic-guided laser can be used for stereotactic neurosurgical procedures including imaging-guided (MRI, CT or angiography) resection of superficial and deep-seated tumours and vascular lesions or endoscopic resection of tumours and cysts in ventricles. Laser- induced interstitial thermotherapy (LITT) is also a minimally invasive neurosurgical approach to the stereotactic treatment of tumours in poorly accessible regions. MRI serves as the most promising technique for on-line monitoring of LITT.

PDT is a relative new technique that has been used for the treatment of brain tumours with HPD, Photofrin or Foscan.

Lasers in Orthopaedics

Patients suffering from herniated discs and unable to recover using physical therapy, can now be treated with lasers. Over 500, 000 Americans undergo low-back pain treatment with a laser each year. Lasers can vaporize tissue in a disc, creating a vacuum.

This causes the disc to shrink away from the pressed nerve, relieving pain. Such surgery eliminates the need for cutting, scarring, hospitalization, postoperative instability, immobility and even general anaesthesia.

Diagnostic Applications of Lasers

'Optical biopsy' or 'optical diagnostics' is a technique whereby light energy is used to obtain information on the structure and function of tissue without disrupting it. Today, this technique employs a number of spectroscopic and imaging methods including absorption, fluorescence, reflectance, elastic scattering and Raman scattering to distinguish malignant from benign tissue, monitor metabolic state and measure local blood flow as well as drug concentration.

The techniques of laser-induced fluorescence (LIF) spectroscopy and imaging are based on the findings that endogenous autofluorescence spectroscopic/imaging patterns differ between normal and premalignant or malignant tissues. For application of LIF to superficial lesions of internal hollow organs optical fibres are inserted through an endoscope, and tissue autofluorescence can be both laser-excited and captured for spectroscopic and imaging analyses. For example, to view mucosal abnormalities or tumours in the trachea and bronchi, lung imaging fluorescence endoscopy (LIFE) uses a fibreoptic endoscope with a blue laser light source to stimulate the natural fluorescence of tissue.

Normal and abnormal tissues respond to this illumination differently, so LIFE images reveal minor abnormalities that would otherwise remain invisible. LIFE can detect early lung cancers or precancerous lesions as small as one millimetre across before they grow into invasive lung cancers. However, due to the complex structure of biological tissue, the low intensity of natural fluorescence signals and the artefacts from scattering and fluorescence reabsorption, photodetection and interpretation of the tissue autofluorescence may be complicated, particularly in the case of inflammatory conditions that may cause false-positive results. These optical diagnostic techniques may be significantly improved by exogenously administered fluorescence compounds or their precursors that selectively localize in specific lesions, such as those photosensitisers used for PDT. ALA- or its derivative-induced endogenously fluorescent PpIX has already shown the enhanced discriminating potential of fluorescence spectroscopy and imaging. Furthermore, the endogenous PpIX can serve as a potent photosensitizing agent for PDT. Most importantly, as a result of rapid advances in methodology and technology of tissue fluorescence imaging and PDT, endoscopists will likely in the coming years have a new armamentarium of diagnostic and therapeutic tools for detection and treatment of superficial premalignancies and malignancies of internal hollow organs during a single procedure. A standard endoscope equipped with an imaging system and a therapeutic illumination system may meet such a requirement for initial photodetection and subsequent PDT of lesions within one treatment session after employing PpIX precursors or other photosensitizers. The combination of endoscopy with the techniques of photodetection and PDT may revolutionize endoscopic technology.

Confocal microscopy is a novel and non-invasive tool that allows for real-time imaging of tissue *in vivo* or fresh biopsies *ex vivo* without the fixing, sectioning and staining necessary for routine histopathology. A confocal microscope consists of a laser light source that illuminates a small spot in tissue. Reflected or fluorescent light from the illuminated spot of the tissue is then imaged through a small pinhole aperture, producing an image of the plane only in focus in the tissue. The confocal microscopy provides fast, high (axial) resolution and high contrast imaging of live tissue that has potential diagnostic applications for lesions.

Optical coherence tomography is a promising diagnostic method that uses low-coherence interferometry to produce a two-dimensional image of optical scattering from internal tissue microstructures in a way that is analogous to ultrasonic pulse-echo imaging. Cross-sectional images of tissue may be obtained *in vivo* with a better spatial resolution than confocal microscopy. This technique is potentially useful for non- invasive diagnosis of the retina and skin tumours with a penetration depth of 0.5–1.5 mm.

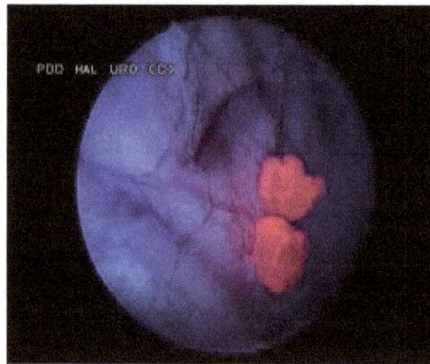

Highly selective ALA hexylester (Hexvix)-induced
PpIX fluorescence in human bladder cancer.

Laser Doppler velocimetry (LDV), also called laser Doppler flowmetry (LDF), is a simple and non-invasive method enabling the monitoring of microvascular blood flow, a very important marker of tissue health. In principle, a monochromatic laser beam is directed at the skin surface. Light that is reflected off stationary tissue undergoes no shift whilst light that is reflected off moving cells (like red blood cells) undergoes Doppler shift. The degree of Doppler shift increases with the velocity of the cells. This light is randomly reflected back from the tissue to a photodetector which calculates the average velocity of cells within the tissue.

As biotechnology evolves, a number of new techniques using a laser such as flow cytometry and mass spectrometry are more often applied to discover new biomarkers for detection of diseases as well as for evaluating the results of treatment.

Laser Nanosurgery in Cell Biology

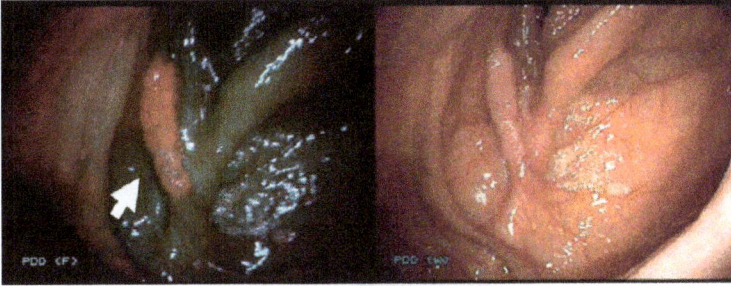

Highly selective ALA hexylester (Hexvix)-induced
PpIX fluorescence (arrow in the left) in human rectal adenoma.

During the past 10 years the quality and availability of pulsed lasers have significantly improved. Very high light intensity can be delivered in an ultrashort period ranging from nanoseconds ($10-9$ s) to femtoseconds ($10-15$ s). By tightly focusing the femtosecond laser pulse with a microscope objective within an extremely small volume, it is possible to achieve ablation on the level of a single cell or a subcellular compartment with a precision in the range of hundreds of nanometres without damaging surrounding structures. Since 1995 when Konig *et al* first demonstrated the dissection of isolated human chromosomes, picosecond UV and femtosecond IR pulsed lasers have increasingly been employed for cellular and subcellular surgery to ablate organelles without affecting cell viability. Figure shows, for example, the entire ablation of a single mitochondrion by a femtosecond laser with precision targeting.

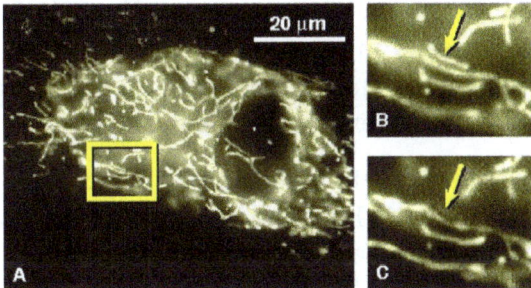

Ablation of single mitochondrion in a cell with a femtosecond pulsed laser. (*a*) Fluorescence image showing multiple mitochondria before the femtosecond pulsed laser application. Ablation of a single mitochondrion before ((*b*), arrow) and after ((*c*), arrow) laser pulses. Note that neighbouring mitochondria are unaffected.

Other subcellular targets include actin filaments and microtubules. Femtosecond lasers have even improved transfection efficiency of DNA in cells for gene therapy. In fact, femtosecond lasers can function as a pair of 'nanoscissors' in subcellular surgery and have potential applications in a single organelle or chromosome dissection, inactivation of specific genomic regions on individual chromosomes and highly localized gene and molecular transfer. The major advantage of pulsed laser nanosurgery is the well-controlled and non-invasive capability of severing subcellular structures with high accuracy in time and three-dimensional space.

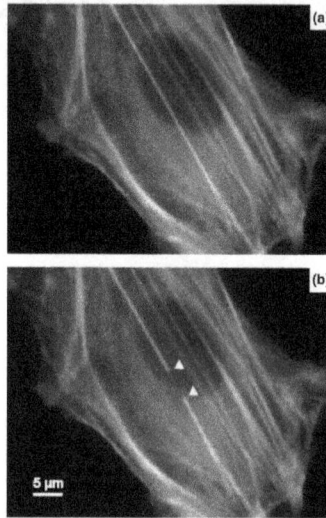

Ablation of actin filaments in a cell with a femtosecond pulsed laser. Fluorescence images of actin filaments before (*a*) and after (*b*) laser dissection (arrowheads) of an actin fibre.

Use of Laser in Entertainment

Scientists have found many tasks and uses for lasers. These devices regularly measure, cut, drill, weld, read, write, send messages, solve crimes, carry telephone conversations, burn plaque out of arteries, and perform delicate eye operations. Over and over again the laser has proved to be an extremely practical tool.

Yet lasers have also proved their usefulness in non-practical applications, especially in the realm of art and entertainment. First and foremost a laser beam is a wand of light; and light itself can be beautiful as well as practical. The sight of a deep red sunset or multicolored rainbow often inspires feelings of happiness, romance, and even awe. For centuries artists have tried to reproduce light's beauty in paintings, and inventors have given artists mechanical tools such as the camera, which uses light to create art that is entertaining (as in the motion picture) as well as beautiful. Because the laser produces a special kind of light, early in the laser era people realized its potential to create special kinds of art and entertainment. And today, lasers are involved in almost all aspects of these fields, from "light shows" to compact discs (CDs) and digital video discs (DVDs), to special effects in the movies.

Combinations of Light and Music

Back in the 1960s when lasers were still relatively new, artists began to use them to produce "light paintings." These took the form of one-time performances in which an artist flashed laser beams in various ways to create visually striking patterns. The beams might be bounced off mirrors placed in preplanned positions or attached to the artist, who would move about, reflecting the rays against walls, glass objects, or into tanks filled with liquid. Another variation involved bouncing the beams against clouds of machine-made fog. Usually the performance was done to music. The effects of these displays could often be exciting to watch, especially at night when the beams glowed brightly against the dark sky.

A huge crowd enjoys a laser show projected onto the granite surface of Stone Mountain near Atlanta, Georgia.

Unfortunately, not many artists could afford the equipment necessary for light paintings. So it became more common for organizations to stage and charge admission fees to see such displays in public performances, which came to be called light shows. The first recorded public laser show took place at Mills College in Oakland, California, on May 9, 1969. Large fairs and celebrations also began to present displays of laser art. The same group that created the Mills College show put on a much more spectacular version in 1970 at the Pepsi-Cola Pavilion at Expo '70 in Osaka, Japan. More than 2 million people attended.

At the Expo '70 show the laser artists set up rotating mirrors and wired them to equipment that played music. They aimed four colored laser beams—red, yellow, green, and blue—at the mirrors. When the music played, the sounds traveled through the wires and caused the mirrors to spin at different speeds; and the mirrors bounced the beams around the room in complex patterns, sometimes to the beat of the music. Many of the spectators reported that the combination of light and music was breathtakingly beautiful.

A much more spectacular display of laser art occurred during the U.S. bicentennial celebration staged at the Washington Monument in 1976. An audience of 4 million people watched the show up close, and the beams could be seen twenty miles away. Other laser artists staged two such large-scale presentations in 1980, one to help celebrate the city of Boston's 350th birthday, the other to enliven a huge party in honor of President Ronald Reagan's inauguration.

In these major laser shows the light beams could be considered the main attraction; the music supported the visual display. But it soon became clear to people in the music business that the reverse could work just as well. Thus it became common to witness laser shows at music concerts, especially rock concerts. In such cases the music performance is the main attraction, while the laser light show takes on the supporting role. Many well-known rock groups and other recording artists have staged these light shows at their concerts; The Who was the first group to do so and Pink Floyd became particularly famous for its laser shows.

From time to time there have been some questions about safety during rock concert laser shows. Some of the early displays allowed the beams to shine into the audience, which was potentially dangerous; the beams are not powerful enough to burn a person's skin, but if a beam shines directly into someone's eye a permanent blind spot can form. Because of this danger a number of countries have established strict rules about how lasers can be used in concerts.

Laser Discs Create a Revolution

While the laser continues to thrill people in large visual shows, it also entertains them on a small scale in their homes. In the late 1970s and early 1980s a revolution in viewing and listening technology began. First came the videodisc player, which plays movies and other shows on a television screen. The disc was encoded with the visual information (the movie) in roughly the same way that computer storage discs are encoded. A laser beam burns patterns into a film that covers the disc and later a small laser inside the player scans the disc and relays the picture to the screen. The picture produced by a videodisc is brighter and sharper than the one produced by a videotape.

Unfortunately, the first videodisc players that came on the market had many problems. A great many had not been built well, and buyers returned them; also, the companies that built them did not make and supply a wide enough variety of movie titles to satisfy customers. Many more titles existed on videotape, and tape players themselves seemed more reliable; so laser videodiscs did not immediately catch on with consumers.

As experts worked to eliminate the bugs from videodisc technology, laser audio discs, which came to be called compact discs, or CDs, hit the market. These did catch on quickly with the public and rapidly replaced traditional long-playing records. One reason for the success of the CD is its excellent sound reproduction. In a phonograph a needle comes in direct contact with the carved grooves on the surface of the record, and the more the record is played the more the grooves wear down. In addition, they can hold only a certain amount of musical information.

CD Player

One side of a compact disc has a reflective coating in which a pattern of pits has been etched. As shown in the enlargement below, a laser beam reflects off these pits onto a light-sensitive transmitter. The transmitter converts the pattern of reflections to electronic signals, which are converted to sound.

In a CD player, by contrast—either audio or video—only a beam of laser light touches the surface of the disc. Barring accidents or misuse, therefore, the discs do not wear out. Moreover, they carry much more information than records and the sound is sharper and more realistic. The other reasons for the success of audio CDs are the same as for the early success of videotapes over videodiscs. The audio CD players proved to be largely reliable, few customers returned them, and manufacturers quickly made tens of thousands of titles available. In fact, many titles appeared on discs that could not be found on traditional records.

Videodisc Comeback and DVDs

Meanwhile, videodisc technology, which had been down but not out, made a comeback. By the early 1990s researchers had considerably improved the technology and produced far more reliable players with disc-produced pictures sharper and more realistic than ever. Furthermore, in an effort to attract new customers, manufacturers offered a wider range of titles and also introduced the concept of deluxe editions of movies. The deluxe versions, which have become almost standard today, often include restored footage (scenes that appeared in the original film but got lost over the years) as well as behind-the-scenes footage of the making of the film, and even outtakes (bloopers).

Later in the 1990s a newer, even more improved version of the laser videodisc—the *d*igital *v*ideo *d*isc, or DVD—appeared. The DVD produces a sharper, more defined picture than either a standard videotape or a laser CD. This is because a videotape picture breaks down into 210 individual horizontal lines, while a CD picture has 425 lines and a DVD picture 540 lines; the more lines, the sharper the image. Expert Rich D'Ambrise explains other superior qualities of DVD:

"DVD discs make CDs look like the 5¼ inch floppy [computer] discs of earlier times. Just a single-sided, single-layer DVD disc offers 4.7 GB [gigabytes, each gigabyte equal to 1,000 megabytes, or MBs] of capacity, which is worlds away from a CD's 680 MB capacity. When we start discussing the 17 GB capacity of a double-sided, dual-layer DVD, it's like comparing the scribbles of a one-year-old toddler to a Monet masterpiece. What makes DVD superior to its CD counterpart is the manufacturing process and internal design. Two injection molds are required to make one DVD, which consists of two banded 0.6 mm discs [as opposed to one in a CD]".

DVD laser technology had become immensely popular in an extremely short time span. By the close of 2001 an estimated 22 million DVD players had been sold; in that same year the film *Shrek* sold a record 2.5 million DVDs, soon surpassed by *How the Grinch Stole Christmas* with sales of 3 million.

Advent of Holographic Images

Another form of laser light–produced art and entertainment is called holography, a special type of photography that creates three-dimensional pictures. By contrast, a standard camera produces pictures that are only two-dimensional. Holography began to develop in the late 1940s, quite a while before lasers appeared. The basic idea was to shine two separate beams of light at a sensitive sheet of photographic film. One beam would bounce off the object being photographed while the other would travel a different path, and both beams would reach the sheet of film at the same time. Once exposed by the light, the film itself became the hologram. Later, when a person shined a third beam at the hologram, a three-dimensional picture of the object was supposed to be visible.

The idea made sense in theory, but it was very difficult to construct a working model. One problem was that the light in the beams had to be coherent, moving along with all the waves in step. Another problem was that both beams had to be monochromatic. Producing two identical beams with these properties was an almost impossible task at that time. Researchers tried all kinds of light sources, but none worked very well and progress in holography was slow all through the 1950s.

Then, in 1960 Theodore Maiman built his ruby laser and holography received a sudden boost. Researchers now had a light source that was bright, coherent, and monochromatic. They found that they could produce two identical laser beams by passing a single laser beam through a device called a beam splitter. These beams bounced off a series of mirrors to reach the photographic film.

Holography

Holograms are photographs that look three dimensional. Objects in a hologram appear to move when viewed from different angles. A hologram is made by directing a laser beam at the object to be photographed. Between the laser and the object, however, is a half-silvered mirror, or beam splitter, which splits the laser beam in two. One of the beams, called the reference beam, is reflected directly from the mirror to the photographic plate. The other, the object beam, first passes through the mirror. Then it reflects off the object and onto the photographic plate. The interference between these two beams when they meet on the photographic plate causes the three-dimensional effect of the hologram.

Almost everyone has seen a hologram at one time or another. The three-dimensional images on credit cards are holograms, as are the many three-dimensional characters and objects portrayed in arcade video games. Sometimes these images seem so real that the spectator invariably reaches out to touch them, only to be reminded that they are illusions.

Some technical problems with holography remain. Objects that are too big cannot be photographed very well. And because monochromatic light must be used, the images produced are in one color. The only way to make multicolored pictures is by combining several different-colored laser beams, which is very difficult to do. The images made this way do not look completely natural. Also, air molecules absorb some of the light and cause the pictures to look grainy. But scientists are working to overcome these problems.

Artists have tried working with holograms but, as with lasers, the equipment is expensive. So holographic art is not yet widespread. A more practical art-related use for holography is in examining ancient paintings. When an old masterpiece is photographed to produce a hologram, experts can detect which sections of the painting are in need of repair.

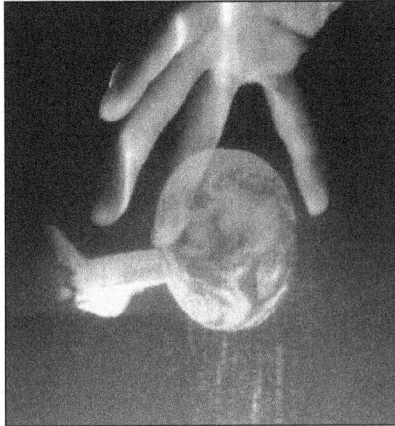

This hologram, a three-dimensional image created by passing laser light through a beam splitter, shows a space shuttle orbiting Earth.

Laser Movie Magic

Still another use for lasers in the entertainment field is the production of special effects for movies. Several companies that produce these effects (usually referred to as "special effects houses") use lasers in highly technical ways to help make their equipment produce truer colors. The first movie to use a laser to print images directly onto the film was Young Sherlock Holmes, released in 1985. Industrial Light and Magic (ILM), perhaps the most famous special effects house, created the effects for the film. In one scene a painted knight on a stained glass church window comes to life. The knight jumps down from the window and chases a priest out of church.

The effect was created in the following way: ILM artists painted the knight onto a TV screen using a special pen that used electricity instead of ink or paint. The image was then stored in a computer that was hooked up to the screen. Next, the artists programmed the computer to rearrange the image so it could be seen from several different angles. Then the computer created pictures of each of the different movements the knight would make in the finished scene. When the artists ordered the computer to play back all these images quickly, the knight appeared to move around on the computer screen.

In the last and most important step the artists transferred the computer images of the knight onto the photographic film. In creating similar effects for previous movies this was done by simply photographing the images directly from the computer screen. But the picture on a computer screen is not as sharp and bright as film-makers would like so the ILM artists decided to connect the computer to a laser. The computer directed the laser to transfer, or "paint," the stored images of the knight right onto the blank film. The knight now showed more detail, and the colors were much more vivid. Later, this film clip of the moving knight was combined with separate film footage of the priest in the church. In the final version that appeared on theater screens the knight seemed to be actually walking around inside the church.

Many later movies have employed this and other laser techniques. The blockbuster Jurassic Park 3, for example, used a laser to create computerized models of the dinosaurs that run amok in the film. First, technicians made a small clay model (called a maquette) of a dinosaur; then they ran a laser beam across the maquette's surface, and the beam transferred highly detailed images of it into a computer. Later, animators used highly sophisticated computer animation programs to make the computer images come to life.

These examples illustrate the use of lasers behind the scenes. But what about lasers on screen? As strange as it might sound, when moviemakers want to portray an actual laser beam on the screen, they cannot use a real laser. For instance, contrary to popular opinion the laser swords used by Luke Skywalker and other characters in the *Star Wars* films were not lasers at all.

There are a number of reasons why such on-screen laser beams have to be faked. First, real laser devices would not produce beams only three feet long; instead, the beams would keep on going and punch holes in the furniture, walls, and bodies of innocent onlookers. Also, the effect of the two beams smashing together like regular metal swords is completely imaginary. Real laser beams would just pass right through each other, an image that would make a movie fight appear disconcerting and more comical than dramatic. The most obvious problem with using real lasers (if such handheld versions could even be built) would be the danger posed by the brightness of the beams. The actors and most of the members of the film crew would all be blind within an hour. For the moment, therefore, the depiction of laser beams on film must be accomplished though more traditional kinds of special effects.

Science Fiction Lasers Perpetuate Misconceptions

Science fiction movies not only utilize lasers in creating their special effects but also regularly depict lasers or laserlike devices. Unfortunately these portrayals of laser light are often inaccurate for the sake of dramatic effect; and this perpetuates several common misconceptions about lasers.

Let's speculate on what would actually happen if you were a member of a large fleet of friendly ships engaging enemy ships in a laser beam battle in space. Science Fiction: You fire a laser weapon at the enemy and you "see" and "hear" the beam emerging from your laser gunports (which violently recoil) and watch the "beam" travel very fast toward the enemy ship, which promptly bursts into flames and explodes very loudly (a sound that you hear instantaneously). Science Fact: You quietly fire an invisible beam from a laser weapon without recoil. Your laser beam travels at the speed of light [so you could not watch it travel]. Sound doesn't travel through a vacuum [so you would hear neither the laser discharge nor the explosion]. Most space ships don't "burst" into flames since there's no air in space to sustain such explosive combustion.

Still, real lasers have added a fresh, visually exciting dimension to the world of entertainment. In the years to come it is certain that scientists and artists will continue to combine their talents to produce many inventive and dramatic new forms of laser-based entertainment.

References

- Lasers-for-lidar-application-parameters-dictate-laser-source-selection-in-lidar-systems: laserfocusworld.com, Retrieved 13 July, 2019

- Laser-surveying-instruments-types-and-uses: digitalhill.com, Retrieved 21 July, 2019

- Laser-communication-system, basic-electronics, projects: electrofriends.com, Retrieved 7 May, 2019

- How-does-laser-cutting-work, education: esabna.com, Retrieved 18 April, 2019

- Faq-how-does-laser-welding-work, faqs, technical-knowledge: twi-global.com, Retrieved 14 July, 2019

- Laser-beam-welding: theweldingmaster.com, Retrieved 11 January, 2019

- Laser-scanning-for-building-design-and-construction: designingbuildings.co.uk, Retrieved 29 June, 2019

- What-is-a-laser-printer-used-for: techwalla.com, Retrieved 30 April, 2019

5

Laser Safety

The safe design, use and implementation of lasers for the purpose of minimizing the risk of accidents related to lasers is termed as laser safety. Some of the different types of hazards related to lasers are electrical hazards, fire hazards and explosive hazards. All the diverse safety principles related to lasers as well as these hazards have been carefully analyzed in this chapter.

Laser beams can be hazardous, particularly for the eye (and sometimes also for the skin), mostly because they can have high optical intensities even after propagation over relatively long distances. Even when the intensity at the entrance of the eye is moderate, laser radiation can be focused by the eye's lens to a small spot on the retina, where it can cause serious permanent damage within fractions of a second – even when the power level is only of the order of a few milliwatts. Damage can result from both thermal and photochemical effects. Laser damage of the eye is not always immediately noticed: it is possible e.g. to burn peripheral regions of the retina, causing blind spots which may be noticed only years later.

Ultraviolet lasers can cause corneal flash burns, a painful condition of the cornea. UV radiation can also cause photokeratitis and cataracts in the eye's lens. (For these reason the XeCl excimer laser has acquired the nickname "cataract machine"). Mid-infrared lasers, particularly those operating at certain wavelengths with very strong absorption in the cornea (e.g. 3 μm or 10 μm), can also cause painful corneal injuries.

Laser Safety Issues

- **Laser pointer, 3 mW:**
  rather bright; could quickly damage the retina, but: blinking reflex helps
- **Small Nd:YAG laser, 100 mW:**
  invisible – no blinking reflex!
  ⇒ rather dangerous for the eyes
- Larger **Nd:YAG laser, 10 W:**
  burns skin and clothes
- Small **Nd:YAG laser** für **Q-switched pulses:**
  very hazardous even for small average output power
- Industrial high power **Nd:YAG or CO_2 laser, 1-10 kW:**
  for welding; not beneficial for skin and eyes!

How much light an eye can tolerate depends on many circumstances: not only the intensity, but also particularly the wavelength and the duration of irradiation (e.g. the pulse duration). There are detailed sets of rules for calculating safe exposure limits (maximum permissible exposure, MPE) for a given situation. Such rules are occasionally revised according to new scientific findings.

Eyes are particularly sensitive, but laser radiation can also cause skin injury. For infrared light, this occurs mainly via thermal effects (thermal skin burns), similar to burning the skin with other means. The penetration depth depends on the wavelength, and for such reasons a laser beam at 1.5 µm wavelength causes more pain on the skin than a 1-µm beam. Whereas such burning should in most cases not have serious long-term consequences, ultraviolet light can in addition induce photochemical reactions. These can lead to changes in the pigmentation, erythema (sunburn), and (most importantly) skin cancer.

Some laser safety issues arise from indirect effects of laser radiation:

- Intense laser beams can incinerate materials and thus possibly start severe fires.

- Particularly in laser material processing, poisonous fumes, dust or hot droplets of molten material can affect nearby workers. For example, fumes containing arsenic, chromium or nickel can occur when metal pieces are cut, and plastic parts (polymers) can form dangerous organic substances.

- Secondary radiation (e.g. ultraviolet light or even X-rays) can be generated when high-intensity beams heat certain targets to high temperatures.

Not only Light is Dangerous

Further issues are not even related to laser beams:

- High electric voltages are used in laser power supplies (e.g. for discharge lamps), creating hazards in maintenance operations or when cables are damaged.

- Some laser systems contain hazardous chemicals such as certain laser dyes.

- Other lasers contain potentially exploding or imploding glass tubes (e.g. arc lamps).

In fact, probably most victims of accidents with lasers have been hurt by such hazards (particularly by electric shocks) rather than by laser radiation.

Particularly Hazardous Situations

The following list of important safety issues can never be complete, but is meant to improve awareness of the multitude of possible hazards:

- High-voltage power supplies can be dangerous if workers can come into contact with the inner parts or with a defective high-voltage cable.

- Some lasers require the handling of hazardous chemicals, e.g. carcinogenic dye solutions in dye lasers. Some of these solutions can penetrate the skin, and therefore need to be handled with special care.

- Near-infrared laser beams are much more hazardous than visible light with the same power level, because their radiation is focused to the retina just in the same way as visible light, whereas the blinking reflex of the human eye (normally closing the eye's lid quickly when the intensity is too high) is not active. Also, no warning is possible e.g. through weak stray light: nothing can be seen when a dangerous beam propagates in an unexpected direction.

- Ultraviolet lasers endanger not only eyes, but also the skin.

- Pulsed laser sources, e.g. Q-switched lasers or regenerative amplifiers, generate pulses with a peak power many orders of magnitude higher than the average output power even of a high-power laser. A single pulse from a hand-held miniature laser can totally destroy an eye.

- In open laser setups, parasitic specular reflections (caused either by parts of the setup or by movable metallic tools, watchbands, rings, etc., but also by the residual reflectivity of anti-reflection coatings) may allow hazardous beams to leave the setup, which might hit someone's eye.

- Optical fibers, e.g. transporting high optical powers between different rooms, may release dangerous radiation when being damaged. They therefore need to be specially protected and marked.

- High-power lasers (e.g. with powers in the kilowatt region) can damage not only the eye but also the skin within short exposure times, and can easily start a fire, e.g. when the beam hits materials such as wood or plastics; toxic fumes may also be generated.

Often less dangerous are the following cases:

- Setups with low-power visible beams, where the blinking reflex of the human eye may provide sufficient protection against occasional exposure of an eye with moderate power levels.

- Sources operating in certain eye-safe spectral regions (e.g. with wavelengths longer than $\approx 1.4\ \mu m$) where the light is absorbed in the eye's lens and therefore cannot reach the (more sensitive) retina.

- Fully closed laser setups with an interlock, which automatically switches off the radiation source as soon as the case is opened.

Safety Classes

To give some guidance on adequate handling and required precautions, laser devices are assigned to different safety classes, with class 1 being the least dangerous (containing e.g. lasers with microwatt power levels) and class 4 the most hazardous one. Note that the assignment to a laser safety class depends not only on the laser power, beam quality and laser wavelength, but also on the accessibility of hazardous areas: even a high-power laser may be in safety class 1 when there is no risk that dangerous radiation can leave a fully encapsulated housing.

Details such as the large diameter or the divergence of involved laser beams are largely ignored in such simplified classification schemes. The concept is solely to classify the laser product itself according to some emission limits, rather than evaluating a particular setup containing a laser. The classification is indirectly based on some exposure limits for the eye, but also takes into account a number of worst case assumptions concerning e.g. the distance of persons from the laser aperture, the exposure duration and the possible use of optical instruments. Therefore, the classification tends to overestimate certain risks, and a complete safety assessment has to consider the details of the whole setup and the way it is used.

Table: International laser safety classes, with somewhat simplified and approximate descriptions. For details, consult the applicable laser safety standard documents.

Safety class	Simplified description
1	The accessible laser radiation is not dangerous under reasonable conditions of use. Examples: 0.2-mW laser diode, fully enclosed 10-W Nd:YAG laser.
1M	The accessible laser radiation is not hazardous, provided that no optical instruments are used, which may e.g. focus the radiation.
2	The accessible laser radiation is limited to the visible spectral range (400–700 nm) and to 1 mW accessible power. Due to the blink reflex, it is not dangerous for the eye in the case of limited exposure (up to 0.25 s). Example: some (but not all) laser pointers.
2M	Same as class 2, but with the additional restriction that no optical instruments may be used. The power may be higher than 1 mW, but the beam diameter in accessible areas is large enough to limit the intensity to levels which are safe for short-time exposure.
3R	The accessible radiation may be dangerous for the eye, but can have at most 5 times the permissible optical power of class 2 (for visible radiation) or class 1 (for other wavelengths).
3B	The accessible radiation may be dangerous for the eye, and under special conditions also for the skin. Diffuse radiation (as e.g. scattered from the some diffuse target) should normally be harmless. Up to 500 mW is permitted in the visible spectral region. Example: 100-mW continuous-wave frequency-doubled Nd:YAG laser.
4	The accessible radiation is very dangerous for the eye and for the skin. Even light from diffuse reflections may be hazardous for the eye. The radiation may cause fire or explosions. Examples: 10-W argon ion laser, 4-kW thin-disk laser in a non-encapsulated setup.

There are different classification schemes (e.g. international and American ones), using classes such as 1 to 4 but with somewhat different definitions. (The American system uses classes I, IA, II, IIIA, IIIB and IV similar to the classes 1 to 4 of the international system, but with significant deviations.) Particularly important standards are:

- The IEC 60825-1 international laser safety standard of the International Electrotechnical Commission (IEC).

- Those based on the US user standard ANSI Z-136 (with various variations Z-136.X, in particular the Z-136.1, revised in 2007).

The IEC standard has been fully adopted by the European standardization organization as EN 60825-1 and is published in national versions such as DIN EN 60825-1 in Germany. Note that these standards cover much more than only defining safety classes; they also determine the measures to be taken in order to work safely with laser products in such classes. There are also government regulations such as the relatively outdated 21 CFR 1040.10, which is still relevant for the US, although the IEC/EN standard is now also accepted there with some additions.

Generally, it is the duty of the manufacturer of a laser product to classify the product and to equip it accordingly with warning labels. However, the classification may change when a laser product is modified by a user, and the user is then responsible for reclassification.

Nominal Hazard Zone

Originally, the required safety measures for a given laser setup where basically determined only by the safety class of the laser. This classification does not reflect details such as beam divergence, which can be very relevant for safety issues: a strongly focused laser beam can be so divergent that within a moderate distance after the focus the intensities fall below the allowable exposure level for the eye. In such situations, one sometimes defines a "Nominal Hazard Zone" (NHZ) within which safe exposure levels may be exceed, in order to apply certain restricting measures to this zone instead of the whole room.

Technical Precautions

Examples of frequently used technical laser safety precautions are:

- The use of protective goggles ($\rightarrow$ *eye protection*), strongly absorbing radiation with wavelengths near the laser wavelength.

- Full or partial encapsulation of laser systems, ideally with absorbing housing materials, avoiding specular reflections.

- Protective housings around dangerous working areas, monitoring the presence of persons e.g. with light curtains, laser scanners or people counters.

- Interlocks that automatically switch off lasers or block laser beams e.g. when a protective box or a door is opened.

- Key-operated switches for power supplies, preventing unauthorized use.

- Written warnings (indicating e.g. The types of lasers behind a door), warning lights (indicating that hazardous laser sources are operated) and automatic door locks, preventing people from entering dangerous areas.

- Beam stoppers (not only for main beams, but also for parasitic reflections), pre-venting dangerous beams from leaving the optical setup.

- Low-power visible pilot beams and the like, marking the paths of dangerous invisible laser beams.

Non-technical Measures

Technical measures alone are generally not sufficient for keeping safety hazards under control. A number of non-technical measures are therefore very important:

- The risks have to be carefully assessed before anything adverse can happen. They need to be reassessed every time when important circumstances change, e.g. the devices used applications, staff, and details of the room.

- On that basis, reasonable ways of dealing with these risks need to be developed. This involves the implementation of technical measures and establishing suit-able working practices. The results need to be clearly described in written safety regulations.

- Adequate safety education needs to be ensured, such that all people who may be at risk are properly informed about both the risks and the proper ways to deal with them. Personal instruction by a knowledgeable person is certainly very valuable, and should be supplemented with additional training materials such as clearly written notes, a laser safety video or DVD.

- All responsibilities need to be properly assigned and clearly defined.

It is also very important to establish a spirit which motivates all staff to take safety is-sues serious, recognize responsibilities for themselves and for their colleagues, suggest practical solutions, etc.

Laser Safety Regulations

Making laser safety regulations for some production facility is a difficult task, and for a re-search laboratory it is even harder. The reason is that there are partially conflicting goals:

- Regulations must be clear and understandable for those reading them.

- The rules should be sensible, ideally under all conceivable circumstances: the implied restrictions should be so that all risks are minimized without being in straight conflict with the actual goals of the work.

- The set of rules should be compact, so that people can be expected to read them carefully and to memorize them.

It is clear that various trade-offs are inevitably involved, e.g. between compactness and suitability for many different circumstances, or between safety and productivity. Giving absolute priority to maximum safety while ignoring productivity and similar practical requirements will not even serve safety, because it increases the risk of regulations being ignored or forgotten. It can be difficult task to analyze existing hazards and to identify the most practical way of dealing with them.

Common Obstacles

Unfortunately, reasonable laser safety regulations are either not in place or (more frequently) routinely ignored in many places such as research and development laboratories. Possible reasons (but not good excuses) are:

- A lack of general knowledge on laser safety issues.

- The lack of available information on specialized safety issues, e.g. Related to potential hazards of ultrashort laser pulses from mode-locked lasers (hazard potential determined only by average power, or also by peak power and pulse duration?), or awareness of risks associated with fumes.

- Unexpected effects such as accidentally misaligned beams, vaporization of poisonous substances, defects or poor design of safety equipment, etc.

- Wrong interpretation of labels like "low-power laser": a 10-mw near-infrared laser may have a low power, but is still very dangerous to the eye.

- Irrational assessment of risks and inappropriate judgment on working routines ("we have always done it like this").

- The general human tendency to underestimate invisible risks, particularly when they occur over long times without apparent effects.

- Missing safety devices (e.g. Insufficient numbers of laser goggles in situations with visitors).

- Highly inconvenient, uncomfortable or otherwise impractical safety devices, e.g. Laser goggles which cannot be used over longer times.

- Excessive pressure to produce results quickly.

- Nonsensical laser safety regulations which undermine the awareness that the adherence to the rules is in the operator's own interest.

- Very formal and abstract sets of rules obviously made primarily for avoiding legal problems for the bosses, rather than providing help in real life.

- The negligence and the bad example of irresponsible supervisors, who sometimes even ridicule more responsible persons.

Due to such factors, which are difficult (if not impossible) to eliminate altogether, perfect laser safety (making accidents impossible) is probably impossible to reach.

However, sensible regulations can greatly diminish the risks without affecting the productivity too severely.

Classes of Laser

Lasers are grouped into seven classes depending on the potential for the beam to cause harm. The hazard and hence the classification depends on the wavelength, power, energy and pulse characteristics. The class of the laser can be used to help decide what safety control measures are required when using the laser. The Accessible Emission Limit (AEL) is the maximum level of laser radiation which a laser can emit (and be accessible) at any time after its manufacture. The AEL depends on the wavelength, exposure duration and the viewing conditions and specifies the maximum output within each laser class.

A distinction should be made between a laser system (which refers to the laser and appropriate energy source) and a laser product (which is defined as "any product or assembly of components which constitutes, incorporates or is intended to incorporate a laser or laser system, and which is not sold to another manufacturer for use as a component (or replacement for such component) of an electronic product"). For example a CD player is a laser product which contains a laser system – the laser diode and power supply. It is the laser product which is placed into one of the seven classes. (The laser product may only consist of a laser system, e.g. a HeNe laser). The descriptions which follow give a summary of the laser classifications.

Class	Basis for Classification
Class 1 Laser Inherently Safe Visible/ non visible	Lasers which are safe under reasonably foreseeable conditions of operation.
Class 1 Laser product Safe as long as not modified	A product that contains a higher class laser system but access to the beam is controlled by engineering means.
Class 2 Low Power Visible only	For lasers, protection of the eyes normally provided by natural aversion blink response which takes approx. 0.25s. Theses lasers are not intrinsically safe. AEL = 1 mW for a CW laser.
Class 1M Safe without viewing aids 302.5 to 4000nm	Safe under reasonably foreseeable conditions of operation. Beams are either highly divergent or collimated but with a large diameter. May be hazardous if user employs optics with the beam.
Class 2M Safe without viewing aids Visible only	Protection of the eyes is normally provided by natural aversion blink response which takes approx. 0.25s. Beams are either highly divergent or collimated but with a large diameter. May be hazardous if user employs optics with the beam.
Class 3R Low/medium power Visible / non-visible	Risk of injury is greater than for the lower classes but not as high as for class 3B. Up to 5 times the AEL for Class 1 and Class 2.

Class 3B Medium / high power Visible / non-visible	Direct intrabeam viewing of these devices is always hazardous. Viewing diffuse reflections is normally safe provided the eye is no closer than 13 cm from the diffusing surface and the exposure duration is less than 10 seconds. AEL = 500mW for a CW laser.
Class 4 High power Visible / non-visible	Direct intrabeam viewing is dangerous. Specular and diffuse reflections are hazardous. Eye, skin and fire hazard. TREAT CLASS 4 WITH CAUTION.

The following labels are associated with the different classes of lasers:

The points of access to areas in which Class 3B or Class 4 lasers are used must be marked with warning signs complying with BS 5378 and the Health and Safety (Safety Signs and Signals) Regulations 1996. The following sign must be used:

Where lasers and laser systems are not adequately labelled (e.g. some American systems have very small labels that are hard to read and do not comply with UK requirements), they must be properly relabelled.

Class 1

Class 1 lasers and laser products are inherently safe under reasonably foreseeable conditions of operation including the use of optical instruments for intra-beam viewing and as such they present no hazard to the eye or skin. They can emit in the visible or invisible. A class 1 laser product may contain a laser system of higher classification, but the engineering controls keep the AEL below the class 1 limit for example a CD/DVD player or a laser printer.

Inherently safe lasers in Class 1 do not need warning labels but lasers which are Class 1 by engineering design and which contain an embedded laser of higher power should be labelled as 'Class 1 Laser Product'. Supplementary information describing the laser product as a 'Totally Enclosed System' with details of the embedded laser clearly displayed may be of value in situation where access to the embedded product is routinely required.

```
CLASS 1 LASER PRODUCT

A TOTALLY ENCLOSED LASER SYSTEM
CONTAINING A CLASS _____ LASER
```

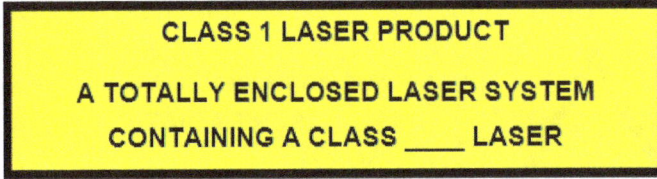

In addition each access panel or protective housing must have a label with the appropriate class inserted and then followed by the hazard warning associated with that class of laser.

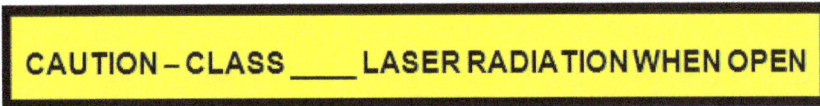

```
CAUTION – CLASS _____ LASER RADIATION WHEN OPEN
```

Class 1M (Safe if not using Viewing Aids)

Class 1M lasers are restricted to 302.5 to 4000nm. They are safe under reasonably foreseeable conditions of operation but may be hazardous if observed using viewing optics. They can be hazardous under two conditions:

a. If the beam is diverging and optics are used within 100mm of the laser aperture to collimate or concentrate the beam into the eye, or

b. If the beam is a large diameter and collimated and optics are used to increase the proportion of the beam that can enter the eye.

Example: Laser diode, LED, fibre communications system.

No hazard warning label is required but there must be an explanatory label bearing the words:

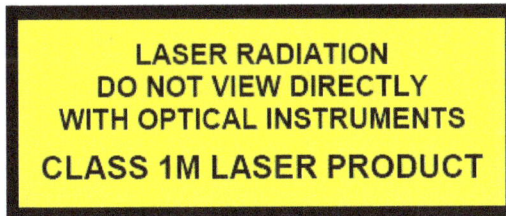

```
LASER RADIATION
DO NOT VIEW DIRECTLY
WITH OPTICAL INSTRUMENTS
CLASS 1M LASER PRODUCT
```

Class 2 (Low Power)

Class 2 only applies to lasers emitting in the visible region, 400 to 700nm. They may be pulsed or continuous wave. Protection of the eyes is normally provided by the aversion response (blinking and/or moving the head) and therefore it is assumed that the exposure time is 0.25s. The AEL for a CW class 2 laser is 1mW.

Example: Alignment HeNe lasers with powers below 1mW, supermarket bar scanners.

A label with hazard warning symbol and an explanatory label as below are required:

LASER RADIATION
DO NOT STARE INTO BEAM
CLASS 2 LASER PRODUCT

Class 2M

Like class 2, class 2M lasers are restricted to the visible range, 400 to 700nm. Protection of the eyes is provided by the aversion response. They may be hazardous if observed using viewing optics under two conditions:

a. If the beam is diverging and optics are used within 100mm of the laser aperture to collimate or concentrate the beam into the eye, or

b. If the beam is a large diameter and collimated and optics are used to increase the proportion of the beam that can enter the eye.

A label with hazard warning symbol and an explanatory label as below are required:

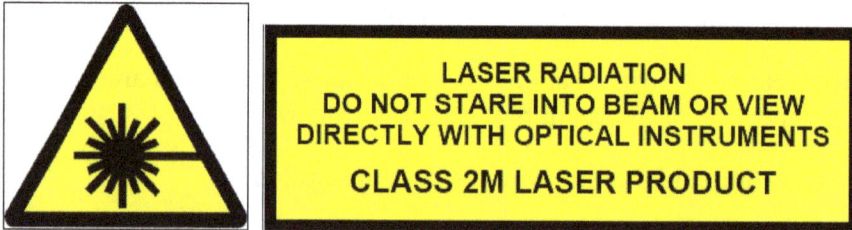

LASER RADIATION
DO NOT STARE INTO BEAM OR VIEW
DIRECTLY WITH OPTICAL INSTRUMENTS
CLASS 2M LASER PRODUCT

Class 3R

Class 3R lasers can have any wavelength between 302.5 and 106nm. The AEL is within five times the AEL of class 2 in the visible (400 to 700nm) and within five times the AEL of class 1 at all other wavelengths. The 'R' refers to 'Relaxed' since this class is a relaxation of the 3B classification.

A label with hazard warning symbol is required for all wavelengths.

For wavelengths 400nm-1400nm ONLY the following explanatory label is needed:

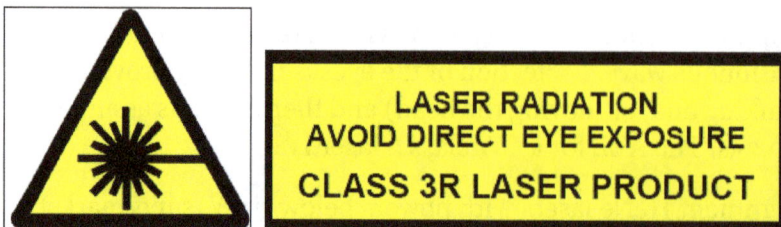

LASER RADIATION
AVOID DIRECT EYE EXPOSURE
CLASS 3R LASER PRODUCT

For other wavelengths the following explanatory label is needed:

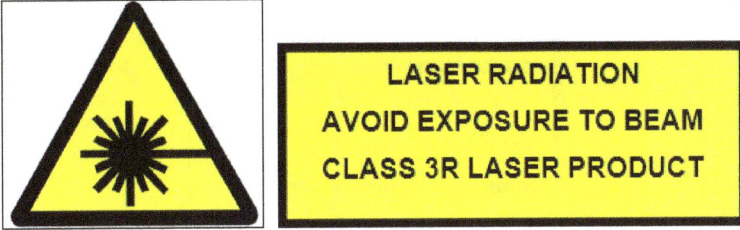

**LASER RADIATION
AVOID EXPOSURE TO BEAM
CLASS 3R LASER PRODUCT**

Class 3B

Class 3B applies to both visible and invisible lasers. Direct intra-beam viewing near these devices is always hazardous. Viewing diffuse reflections is normally safe provided the eye is no closer than 13cm from the diffusing surface and the exposure duration is less than 10s.

A label with hazard warning symbol and an explanatory label as below are required:

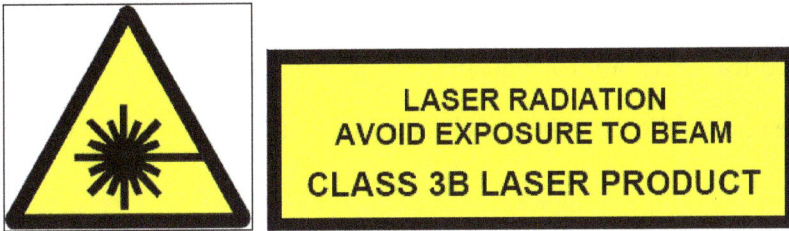

**LASER RADIATION
AVOID EXPOSURE TO BEAM
CLASS 3B LASER PRODUCT**

Class 4

Class 4 lasers are hazardous for direct intra-beam, direct reflected and diffuse reflected beam viewing. They may cause eye or skin injuries and also may constitute a fire hazard.

A label with hazard warning symbol and an explanatory label as below are required:

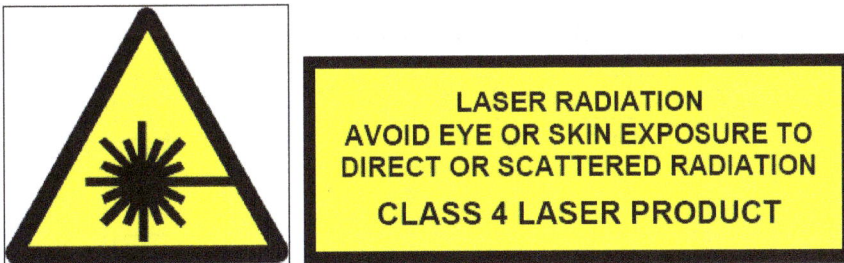

**LASER RADIATION
AVOID EYE OR SKIN EXPOSURE TO
DIRECT OR SCATTERED RADIATION
CLASS 4 LASER PRODUCT**

Aperture Labels for Class 3R, Class 3B and Class 4 lasers

Each Class 3R, Class 3B and Class 4 laser product must display a label close to where the beam is emitted bearing the words 'Laser Aperture' or 'Avoid Exposure - Laser Radiation is Emitted from this Aperture'. This label can take the form of an arrow if this displays more meaning:

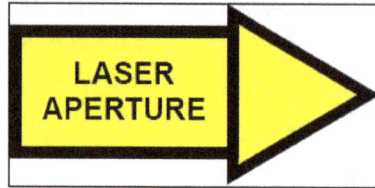

Radiation Output and Standards Information

All laser products, except for low power Class 1 devices, must describe the following details on an explanatory label:

- Maximum output;

- Emitted wavelength;

- Whether laser beam is visible, invisible or both;

- Pulse duration (if appropriate);

- Name and publication date of classification standard.

Maximum Permissible Limit

Maximum permissible exposure (MPE) is the highest power or energy density of the light source measured in W cm^{-2} or J cm^{-2}, respectively, that is considered safe and has negligible probability of causing any damage to the eyes. MPE is usually taken as 10% of the power or energy density that has 50% probability of causing damage under worst-case conditions.

Table: Pathological effects of different wavelength ranges.

Wavelength range	Pathological effect
180–315 nm	Photokeratitis: inflammation of cornea, equivalent to sun burn
315–400 nm	Photochemical cataract
400–780 nm	Photochemical damage to retina, retinal burn
780–1400 nm	Cataract, retinal burn
1.4–3.0 μ m	Aqueous flare, cataract, corneal burn
3.0–1000 μ m	Corneal burn

Table: Laser safety standard old system.

Safety class	Description
I	A class I laser is safe and there is no possibility of eye damage. This can be either due to low output power with no risk of eye damage after hours of exposure, or due to the laser being contained inside an enclosure such as in the case of a compact disk player or laser printer.
II	This class refers only to lasers emitting in the visible spectrum with output power up to 1.0 mW. The lasers are safe due to the blink action of the eye unless deliberately staring into the beam for an extended period of time. Most laser pointers belong to this category.
IIa	For continuous exposure for a period of >1000 s, lasers at the low power end of the class II category may produce retinal burn.
IIIa	Lasers with power level >1.0 mW and <5.0 mW and power density <2.5 mW cm^{-2} belong to class IIIa. These lasers are dangerous when used in combination with optical instruments and also dangerous to the naked eye for direct viewing for more than 2.0 min.
IIIb	Lasers with power level 5.0–500 mW belong to Class IIIb. Direct viewing of these lasers may cause damage to the eye. A diffuse reflection is generally not hazardous but both direct viewing and specular reflection are equally dangerous. Class IIIb lasers, which are towards the high power end, may present a fire hazard or cause skin burn.
IV	Lasers belonging to class IV have power levels >500 mW. Class IV lasers can cause permanent eye damage or skin burn even when used without optical instrumentation. Diffuse reflection from these lasers can also be hazardous to eyes or skin within the nominal ocular hazard zone. Many industrial, medical, scientific and military lasers are of the class IV category.

Table: Laser safety standard: Revised system.

Safety class	Description
1	The accessible laser radiation is not dangerous under reasonable conditions of use. A class 1 laser is safe under all conditions of normal use. This implies that while viewing a class 1 laser with the naked eye or with typical magnifying optics such as a telescope or a microscope; the MPE limit cannot be exceeded.
1M	The accessible laser radiation is not hazardous, provided that no optical instruments are used which may, for example, focus the radiation. A class 1M laser is safe for all conditions except when passed through magnifying optics such as telescopes and microscopes. The classification is applicable to lasers with power level greater than the limit specified for class 1 lasers provided the laser energy entering the pupil of the eye does not exceed the limits of class 1 lasers due to the large divergence of the laser.
2	The accessible laser radiation is limited to the visible spectral range (400–700 nm) and to 1 mW accessible power. Due to the blink reflex action of the eye, it is not considered dangerous for limited exposure up to 0.25 s.
2M	A class 2M laser is also safe because of the reflex action of the eye, with the additional restriction that it is not viewed through optical instruments. As for class 1M lasers, the classification is also applicable to laser beams with power level >1 mW provided the beam divergence is large enough to prevent laser energy passing through the pupil of the eye to exceed the limits of class 2 lasers.

3R	The accessible radiation may be dangerous for the eye, but can have at most 5 times the permissible optical power of class 2 lasers emitting in visible spectrum and class 1 for other wavelengths. The MPE can be exceeded with class 2M lasers with a low risk of injury.
3B	The accessible radiation may be dangerous for the eye and, under particular conditions, also for the skin. Diffuse radiation scattered from a diffuse target is normally harmless. The accessible emission limit is 500 mW for CW visible lasers emitting in visible spectrum and 30 mW for pulsed lasers. In the case of direct viewing of class 3B lasers, protective eye wear is required.
4	The accessible radiation of a class 4 laser is very dangerous for the eye and for the skin. Light from diffuse reflections and indirect viewing may be hazardous for the eye. Class 4 lasers must be equipped with a key switch and have an in-built safety interlock. Most industrial, medical, scientific and military lasers belong to this category.

Hazard Zone and Individual Protection

This is the distance from the source at which the intensity or the energy per surface unit becomes lower than the Maximum Permissible Exposure (M.P.E.) on the cornea and on the skin. The laser beam can thus be considered as dangerous if the operator is closer from the source than the N.O.H.D.

Like the M.P.E., this distance depends on several parameters:

- The beam characteristics: Output power, diameter and divergence.

- The M.P.E. value on the cornea.

- Eventually, the optical system inserted in the beam trajectory.

For example, this distance can be extremely long for class 3B and 4 laser sources. It is thus necessary to stop the beam at the end of the optical system. When looking at the beam with an optical system, one has to consider the possible higher intensity entering the eye, and thus to expand the evaluated N.O.H.D. (called afterwards expanded N.O.H.D.)

As long as the beam propagates freely, this distance can be evaluated according to the following expression:

$$N.O.H.D = \frac{1}{\theta} \sqrt{\frac{4 \cdot P_0}{\pi \cdot M.P.E} - (2 \cdot w)^2}$$

In this formula, N.O.H.D is the Nominal Ocular Hazard Distance (in meter), P_o the power of the source (in Watts) or eventually the total energy carried by one pulse (in Joules), M.P.E the Maximum Permissible Exposure (in W/rad or J/m²) , w the waist of the Gaussian beam (m), and θ the divergence of the beam.

When using an optical system to look at the beam, one has to take into account the beam focusing induced by the system. Defining f the focal length f of the optical system and α the half-aperture angle of the beam, the expression turns to:

$$N.O.H.D = f + \frac{1}{\tan(\alpha)} \sqrt{\frac{P_0}{\pi.M.P.E.}}$$

Nominal Ocular Hazard Area (N.O.H.A.)

Inside this area, the intensity or the energy per surface unit is higher than the M.P.E. on the cornea. The size of this area is defined by the N.O.H.D. However, it is very difficult to define this area as it depends on the environment (dusty or not) and on the objects than can be on the beam trajectory – in other words, one has to take into account the specular reflections.

Optical Density of the Eyewear

Protective eyewear in the form of spectacles or goggles with appropriately filtering optics can protect the eyes from the reflected or scattered laser light with a hazardous beam power, as well as from direct exposure to a laser beam. Eyewear must be selected for the specific type of laser, to block or attenuate in the appropriate wavelength range. For example, eyewear absorbing 532 nm typically has an orange appearance, transmitting wavelengths larger than 550 nm. Such eyewear would be useless as protection against a laser emitting at 800 nm. Eyewear is rated for optical density (OD), which is the base-10 logarithm of the attenuation factor by which the optical filter reduces beam power. For example, eyewear with OD 3 will reduce the beam power in the specified wavelength range by a factor of 1,000. In addition to an optical density sufficient to reduce beam power to below the maximum permissible exposure, laser eyewear used where direct beam exposure is possible should be able to withstand a direct hit from the laser beam without breaking for at least 10 seconds. The protective specifications (wavelengths and optical densities) are usually printed on the goggles, generally near the top of the unit.

Therefore, in selecting protective eyewear two characteristics must be considered:

- Optical density;
- Damage threshold.

Example: Determine the required optical density of eyewear for working with a 0.5 W laser that emits 532nm light.

Solution: The limiting aperture for visible light is 0.7 cm corresponding to an area of 0.385 cm².

Therefore, the irradiance (E) on the eye as defined by the limiting aperture is:

$$E = 0.5 \text{ W}/0.385 \text{ cm}^2$$

$$E = 1.30 \text{ W/cm}^2$$

The transmitted irradiance must be no greater than the MPE for this wavelength, i.e. 2.5×10^{-3} W/cm². Therefore, the required optical density of the protective eyewear is:

$$D_\lambda = \log (1.30/2.5 \times 10^{-3})$$

$$D_\lambda = 2.7$$

The optical density of protective eyewear depends on the wavelength of the incident light. While most protective eyewear offers protection over a range of wavelengths, not all of the wavelengths will be attenuated to the same extent. Therefore, in selecting protective eyewear it is important to ensure that the optical density of the eyewear is adequate for the wavelength of interest.

Studies have shown that protective filters can exhibit non-linear effects such as saturable absorption when the filter is exposed to pulses of ultra-short duration (i.e. $< 10^{-12}$ s). Therefore, the optical density of the filter may be considerably less than expected for very short pulses and it is strongly recommended that the manufacturer be consulted when choosing eyewear for these types of lasers.

The other factor that is important in selecting protective eyewear is the damage threshold specified by the manufacturer. The damage threshold is the level of irradiance above which damage to the filter will occur from thermal effects after a specified period of time - usually 10 seconds. Once the damage threshold is exceeded, the filter ceases to offer any protection from the laser radiation and serious injury can result. The damage threshold varies with the type of material used in the filter and some typical ranges are given below:

Type of Material	Damage Threshold (W/cm²)
Plastic	1 - 100
Glass	100 - 500
Coated Glass	500 - 1,000

Intense, Q-switched, laser pulses can cause filters to crack and shatter up to 30 minutes following the exposure and some filters have exhibited photo bleaching after exposure to Q-switched laser pulses.

Other factors that should be considered when selecting protective eyewear include:

- The need for side-shield protection and peripheral vision.
- Prescription eyewear.
- Comfort and fit.
- Strength and resistance to mechanical trauma and shock.
- Potential for producing specular reflections off of the eyewear.
- Need for anti-fogging design or coatings.
- The requirement for adequate visible light transmission.

Protective eyewear must be clearly labelled with the optical density and wavelength for which protection is provided. In a multi-laser environment color coding of the protective eyewear is recommended.

Protective eyewear must be regularly cleaned and inspected for pitting, crazing, cracking, discoloration, mechanical integrity, the presence of light leaks or coating damage. When damage is suspected the protective eyewear should be either retested for acceptability or discarded.

When purchasing protective eyewear the wavelength, optical density, damage threshold, shelf life, storage conditions and limitations for use should be requested from the manufacturer before the purchase is made. This will ensure that the eyewear is adequate for the anticipated conditions of use.

Injuries Related to Laser use on Skin

Skin is the largest organ of the body and, as such, is at the greatest risk for coming in contact with the laser beam. The most likely skin surfaces to be exposed to the beam are the hands, head, or arms.

Lasers can harm the skin via photochemical or thermal burns. Depending on the wavelength, the beam may penetrate both the epidermis and the dermis. The epidermis is the outermost living layer of skin. Far and Mid-ultraviolet (the actinic UV) are absorbed by the epidermis. A sunburn (reddening and blistering) may result from short-term exposure to the beam. UV exposure is also associated with an increased risk of developing skin cancer and premature aging (wrinkles, etc) of the skin.

Skin Components

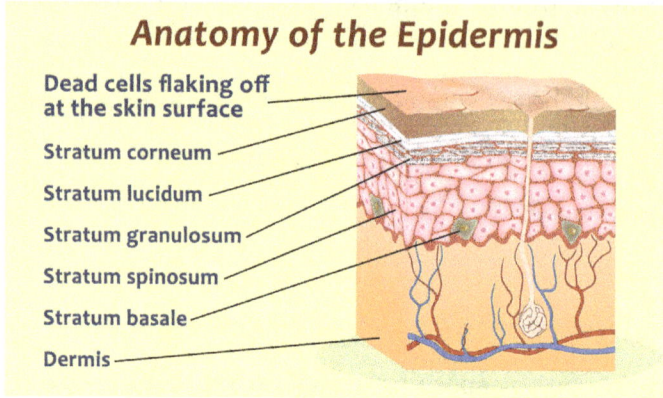

Anatomy of the Epidermis

Dead cells flaking off at the skin surface
Stratum corneum
Stratum lucidum
Stratum granulosum
Stratum spinosum
Stratum basale
Dermis

- Stratum Corneum (Dead Layer): The stratum corneum is the outermost layer or the horny layer. It consists of flattened, dead epidermal cells that protect the living tissue from water loss, injury from physical objects, and radiant energy. The thickness is 8-20 um except on the soles of the feet and the palms of the hands where it is 500-600 μm.

- Epidermis: The epidermis is the outermost layer of living tissue, where the tanning process takes place. It has a relatively uniform thickness of about 50-150 μm.

- Dermis Corium: The dermis corium is made up largely of connective tissue which gives the skin its elasticity and supportive strength. Included in this layer are nerve cells, blood vessels, and lymphatic glands. The thickness of this layer varies over the body from 1-4 mm.

- Subcutaneous Tissue: The subcutaneous tissue is made up mostly of fatty tissue serving as insulation and as a shock absorption medium. The thickness of this layer varies according to the area of the body as well as from person to person.

How does Laser Light affect the Skin?

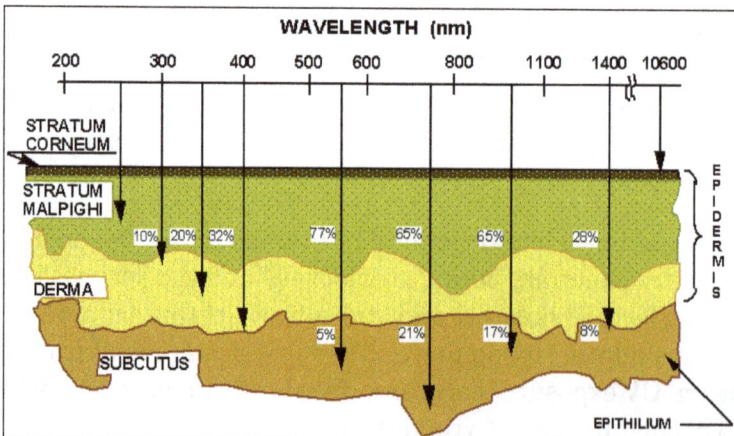

As indicated by the illustration above, different wavelengths of light penetrate the skin in different ways. At approximately 750 nm, absorption to the subcutis occurs.

Laser effects on tissue depend on - the power density of the incident beam, absorption of tissues at the incident wavelength, time beam is held on tissue, and the effects of blood circulation and heat conduction in the effected area.

Tissue Damage from a CO_2 Laser

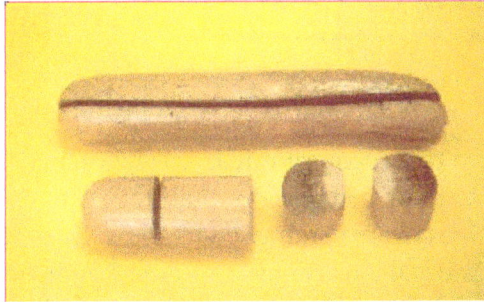

250 Watt Laser Moving at 1 Inch per Second.

250 Watt Laser in Single Pulses.

Immediate Effects

As shown above, the immediate effect of exposure to laser light above the biological damage threshold is normally burning of the tissue. Injury to the skin can result either from thermal injury following temperature elevation in skin tissues or from a photo-chemical effect (e.g., "sunburn") from excessive levels of actinic ultraviolet radiation.

Some individuals are photosensitive or may be taking prescription drugs that induce photosensitivity. Particular attention must be given to the effect of these (prescribed) drugs, including some antibiotics and fungicides, on the individual taking the medication and working with or around lasers.

Delayed Effects

The possibility of adverse effects from repeated or chronic laser irradiation to the skin has been suggested, although it is normally discounted. Only optical radiation in the ultraviolet region of the spectrum has been shown to cause long-term, delayed effects. These effects are: accelerated skin aging and skin cancer. At present, laser safety standards for exposure of the skin attempt to take these adverse effects into account.

Hazards Related to Laser

Electrical Hazards

The use of lasers or laser systems can present an electric shock hazard. These exposures can occur during laser set up or installation, maintenance, modification, and service, where equipment protective covers is removed to allow access to active components. A contact with energized electrical conductors contained in device control systems, power supplies, and other components is another way to receive an electric shock. With the use of large power supplies and repetitively pulsed lasers, there is a great potential for electric shock. Shocks usually happen when a person is working on equipment that is not properly grounded or has a large capacitor bank that was not discharged. Most injuries to personnel involving lasers are of this type.

Electric shock is a very serious opportunistic hazard where the occurrence and outcome are difficult to predict, and loss of life has occurred during electrical servicing and testing of laser equipment incorporating high voltage power supplies.

Protection against accidental contact with energized conductors by means of a barrier system is the primary methodology to prevent electric shock accidents.

The frames, enclosures and other accessible non-current-carrying metallic parts of laser equipment should be grounded. Grounding should be accomplished by providing a reliable, continuous metallic connection between the part(s) to be grounded. A presence of "Emergency Power Off" switch will allow the elimination of electrical hazards during emergencies.

Preventive Measures

- Fluids should not be used or placed near the laser system.
- The laser system should be labeled with the electrical rating, frequency and watts.
- Proper grounding should be used for metal parts of the laser system.
- Assume that all floors are conductive when working with high voltage.
- Consider safety devices such as appropriate rubber gloves and insulating mats.

- Make sure that the combustible components of the electrical circuit are short circuit tested.

- Check that each capacitor is discharged, shorted and grounded before allowing access to the capacitor area.

- Inspect capacitors containers for deformities or leaks.

- Avoid wearing rings, metallic watchbands and other metallic objects when working near high voltage environment.

- Prevent explosions in filament lamps and high pressure arc lamps.

- Inspect regularly the integrity of electrical cords, plugs, and foot pedals.

- Only qualified persons authorized to perform service activities should access a laser's internal components.

- Do not work alone.

- When possible, only use one hand when working on a circuit.

- Follow lockout/tag-out procedures when applicable.

Fire Hazards

A fire can occur when a laser beam (direct or reflected) strikes a combustible material such as paper products, plastic, rubber, human tissues, human hair, and skin treated with acetone and alcohol-based preparations. The risk of fire is much greater in oxygen-rich atmosphere.

The three components required for a fire to start are:

- A combustible material;

- An oxidizing agent;

- A source of ignition.

Therefore, to reduce the risk of fire in laser applications, great care must be taken to keep these components physically separated from each other. In general, Class 3B lasers do not pose a fire hazard, while Class 4 lasers do.

Enclosure of Class 4 laser beams can result in potential fire hazards if enclosure materials are likely to be exposed to irradiances exceeding 10 W/cm² or beam powers exceeding 0.5 W. Under some situations where flammable compounds or substances exist, it is possible that fires can be initiated by Class 3B lasers.

Opaque laser barriers (e.g., curtains) can be used to block the laser beam from exiting the work area during certain operations. While these barriers can be designed to offer a range of protection, they normally cannot withstand high irradiance levels for more than a few seconds without some damage (e.g., production of smoke, open fire, or penetration). Users of commercially available laser barriers should obtain appropriate fire prevention information from the manufacturer. Users can also refer to NFPA115 (Standard for Laser Fire Protection) for further information on controlling laser induced fires. The use of flame retardant materials is encouraged wherever applicable.

Operators of Class 4 lasers should be aware of the ability of unprotected wire insulation and plastic tubing to catch on fire from intense reflected or scattered beams, particularly from lasers operating at invisible wavelengths.

Barriers such as black photographic cloth or black paper products are used in a wide variety of applications for the purpose of containing the beam. These materials must not be used as the primary barrier for a high-powered Class 4 system. Beams of sufficient energy will burn this material quickly, causing smoke, fire, and breach of the barrier. The use of beam blocks and beam stops is highly encouraged in such situations.

Prevention Measures

- Maintain precise control of the laser beam.

- Eliminate surfaces that can reflect laser beam.

- Do not use combustible materials as enclosure material with Class 4 lasers or stored nearby.

- Be prepared – have appropriate fire extinguisher in close proximity and easily accessible.

Explosive Hazards

Many metals or their oxides can represent fire or explosion hazards under certain circumstances, such as exposure to air, moisture, water, chemicals, shocks and impacts. Examples are alkali metals, aluminum, magnesium, titanium, hafnium, plutonium, thorium, uranium, and zirconium. Extreme care should be taken during transfer and dispensing powders, in order to prevent the formation of dust or the introduction of oxygen.

High-pressure arc lamps, filament lamps, and capacitor banks in laser equipment should be enclosed in housings that can withstand the maximum explosive pressure resulting from component disintegration. The laser target and elements of the optical train that may shatter during laser operation should also be enclosed to prevent injury to operators and observers. Explosive reactions of chemical laser reactants or other laser gases may be a concern in some cases. There have been several reports of explosions caused by the ignition of dust that has collected in ventilation systems serving laser processes. The potential for such can be greatly minimized by good maintenance practice.

In addition, explosions can be caused by the beam from a Class 4 laser hitting a gas cylinder, regulator, or delivery hose.

References

- Laser-safety: rp-photonics.com, Retrieved 1 August , 2019
- Appendix1classification, lasers, guidance, healthsafetywellbeing, services: warwick.ac.uk, Retrieved 9 May, 2019
- Non-beam-hazards, laser-safety, radiation-protection, documents, resource: ehs.lbl.gov, Retrieved 8 August, 2019
- Laser-biological-hazards-skin, training, laser: oregonstate.edu, Retrieved 31 March, 2019
- Laser-safety-v4, docs, assets, chemistry, science: uvic.ca, Retrieved 14 July, 2019

Permissions

All chapters in this book are published with permission under the Creative Commons Attribution Share Alike License or equivalent. Every chapter published in this book has been scrutinized by our experts. Their significance has been extensively debated. The topics covered herein carry significant information for a comprehensive understanding. They may even be implemented as practical applications or may be referred to as a beginning point for further studies.

We would like to thank the editorial team for lending their expertise to make the book truly unique. They have played a crucial role in the development of this book. Without their invaluable contributions this book wouldn't have been possible. They have made vital efforts to compile up to date information on the varied aspects of this subject to make this book a valuable addition to the collection of many professionals and students.

This book was conceptualized with the vision of imparting up-to-date and integrated information in this field. To ensure the same, a matchless editorial board was set up. Every individual on the board went through rigorous rounds of assessment to prove their worth. After which they invested a large part of their time researching and compiling the most relevant data for our readers.

The editorial board has been involved in producing this book since its inception. They have spent rigorous hours researching and exploring the diverse topics which have resulted in the successful publishing of this book. They have passed on their knowledge of decades through this book. To expedite this challenging task, the publisher supported the team at every step. A small team of assistant editors was also appointed to further simplify the editing procedure and attain best results for the readers.

Apart from the editorial board, the designing team has also invested a significant amount of their time in understanding the subject and creating the most relevant covers. They scrutinized every image to scout for the most suitable representation of the subject and create an appropriate cover for the book.

The publishing team has been an ardent support to the editorial, designing and production team. Their endless efforts to recruit the best for this project, has resulted in the accomplishment of this book. They are a veteran in the field of academics and their pool of knowledge is as vast as their experience in printing. Their expertise and guidance has proved useful at every step. Their uncompromising quality standards have made this book an exceptional effort. Their encouragement from time to time has been an inspiration for everyone.

The publisher and the editorial board hope that this book will prove to be a valuable piece of knowledge for students, practitioners and scholars across the globe.

Index